PROJECT ANALYSIS & CAS

SYSTEMS ANALYSIS AND DESIGN

A PROJECT APPROACH

DAVID HARRIS
College of the Redwoods

The Dryden Press
Harcourt Brace College Publishers
Fort Worth Philadelphia San Diego New York Orlando Austin San Antonio
Toronto Montreal London Sydney Tokyo

Cover photograph: Dick Patrick

Address for Editorial Correspondence
The Dryden Press, 301 Commerce Street, Suite 3700, Fort Worth, TX 76102

Address for Orders
The Dryden Press, 6277 Sea Harbor Drive, Orlando, FL 32887
1-800-782-4479, or 1-800-433-0001 (in Florida)

ISBN: 0-03-011618-x

Printed in the United States of America

4 5 6 7 8 9 0 1 2 3 023 9 8 7 6 5 4 3 2 1

The Dryden Press
Harcourt Brace College Publishers

TABLE OF CONTENTS

PART 1 - INTRODUCING THE TEAM PROJECT

PART II - COMPLETING THE TEAM PROJECT: STEP BY STEP

PART I

INTRODUCING THE TEAM PROJECT

This part of the casebook concerns the team project. It begins with a general description of the type of systems projects you will be assigned to work on. This is followed by a detailed discussion of the project work you will be expected to complete as a member of a team. Appropriately, some specific suggestions are offered to help you and your team work effectively. Several sample project management worksheets are included to help you organize the considerable amount of project related documents and computer files that you will create during this exercise.

Four project packets appear at the end of Chapter 2. Your team may be assigned one of these projects or a project of similar complexity. You may find it useful to read all four of these brief packets to help you identify information system issues that are common to small enterprises.

CHAPTER 1

THE TEAM PROJECT

TEAM PROJECT GUIDELINES

The Goal: The purpose of the team project is to provide you with an opportunity to learn systems analysis, design, development, and implementation within a controlled environment. The emphasis is on producing quality microcomputer-based information systems for small enterprises.

The Team: Undoubtedly, each team will include students with different computer training and work experience. Although every student should have completed an introductory computer course and be somewhat familiar with database fundamentals, each team should have at least one member with some programming skills.

The Budget: To minimize distractions that accompany complex hardware and software implementations, the delivery budget for your project is $10,000. Your team is to assume a billing fee of $50.00 per "team" hour, that is, if four members devote a total of 24 hours to a project task, the team should only charge 6 labor hours to the project. The project will run for 12 to 16 weeks, depending on your school's academic calendar. Assuming that your team bills an average of 5 or 6 hours per week, this leaves well over one-half of the budget for computer hardware, software, supplies, and incidentals.

The Small Enterprise: Although your project is fictitious, it is based on a real-life small enterprise. In general, these enterprises can be expected to have manual information systems in place, with only limited computer support, and, while the enterprise personnel are computer literate, they remain apprehensive about computer technology.

The Project: Your project will involve one of many different types of small enterprises—profit, non-profit, service, health, manufacturing, agriculture, finance, entertainment, construction, recreation, retail, travel, real estate, automotive, legal, public, private, and so on. The intent is to demonstrate that universal principles apply to information system projects, regardless of the particular enterprise characteristics. Thus, every project presents the same set of challenges, summarized as follows:

1. A transaction processing system
2. A client or customer database maintenance
3. A budgeting or forecasting spreadsheet application
4. A standard correspondence and mail-merge operation

The Process: To begin the process, your team receives a project packet containing a general description of the small enterprise, a statement describing the user's information system request, and a collection of sample documents relevant to the project. Through a series of interviews and correspondence, your instructor, acting as the agent for the enterprise, will clarify the information system requirements.

There are two major portions to your project: the first involves systems analysis and design, concluding with a team presentation and proposed system design report; the second requires you to develop and implement the agreed upon design. Along the way, a system prototype presentation and report provides for active user participation in the project. A user training session, accompanied by product documentation and a final report, completes the second portion of the project.

The Lab: Throughout this process, you will create several traditional system models and project management documents (charts, diagrams, project dictionaries, status reports, etc.). To the extent possible, automated project support software will be available to make this work easier. With minor exceptions, you will have full access to the hardware and software resources of the computer lab. Of course, you will be constrained by two practical considerations: your school's computer lab resources, and the time required to learn how to use new products.

TEAM PROJECT PRESENTATIONS AND DELIVERABLES

Your computer information system project includes several milestone events, which mark the progress of your work. To help you schedule your work and plan for these events, Figure 1-1 presents two weekly time lines. One is annotated with the project presentations and product deliveries that correspond to the 16-week Cornucopia sample project in Appendix B. A second, blank time line, is for your use. The remainder of this section offers detailed outlines for the project presentations and deliverables.

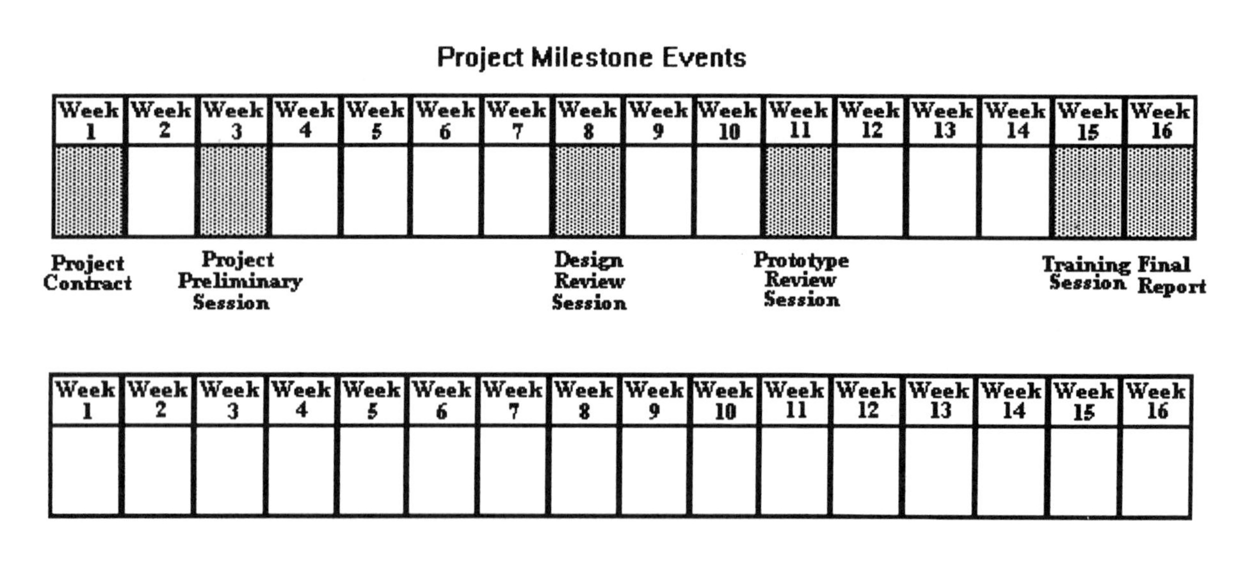

Figure 1-1

Project Contract

Written Report:

1. Cover letter
2. Project Contract
    a. Problem summary
    b. Project scope
    c. Project constraints (cost, delivery date)
    d. Objectives

Project Preliminary Session

Written Report:

1. Cover letter, title page, table of contents
2. Overview of proposed new system (USD)
3. Proposed project timetable (project status report)
4. Proposed project cost breakdown (project budget)

Oral Presentation:

1. Introduction of team members (1 min.)
2. Distribute handouts for items 2-4 above
3. Discuss summary of project requirements (9 min.)
4. Discuss visuals for items 2-4 above (10 min.)
5. Question and answer period (5 min.)

Design Review Session

Written Report:

1. Cover letter, title page, table of contents
2. Overview of the new system design
    DFD, ERD, USD, menus, project dictionary
    system and subsystem flowcharts
    input/output design layouts
    interactive design layouts
3. Hardware, software specifications and cost estimates
4. Cost/Benefit analysis and rationale
5. Project status report
6. Project budget

Oral Presentation:

1. Discuss purpose of the session (1 min.)
2. Distribute handouts for items 2-4 above
3. Discuss visuals for items 2-4 above (24 min.)
4. Question and answer period (5 min.)

Prototype Review Session

Written Report:
1. Cover letter, title page, table of contents
2. USD with prototyped segments highlighted
3. Updated new system design
   - revised DFD, ERD, USD, menus, project dictionary
   - GUID screen images
   - sample reports
   - modular program structure charts
4. Updated project status report
5. Updated project budget

Oral Presentation and Demonstration:
1. Discuss the prototyping methodology (5 min.)
2. Distribute handouts for items 2-3 above
3. Discuss visuals for items 2-3 above (5 min.)
4. Demonstrate prototyped segments (20 min.)
5. Question and answer period (10 min.)

Training Session

Written Report:
1. Cover letter, title page, table of contents
2. Description of training objectives
3. User training session reference documents

Oral Presentation and Hands-On Exercise:
1. Discuss purpose of the session (1 min.)
2. Distribute handouts for items 2-3 above
3. Brief demonstration of the system (4 min.)
4. Hands-on user training session (25 min.)
5. Question and answer period (10 min.)

Final Project Report and Documentation

Written Report:
1. Cover letter, title page, table of contents
2. Final project status report with an analysis
3. Final project budget with an analysis
4. Narrative of proposed system enhancements

Project Documentation:
1. Training Manual
2. Procedures Manual
3. Reference Manual
4. Floppy disk with all project files

TEAM PROJECT ASSIGNMENTS

The team project assignments are designed to help your team complete the project deliverables and prepare for the milestone events detailed above. For easy reference, this section begins with a summary of the products you must submit for each assignment. The summary is followed by the detailed assignments.

Team Assignment Summary

Assignment 1 - team name, logo, letterhead, advertising flyer

Assignment 2 - project questions*, initial response to client

Note: The PROJECT CONTRACT is due at this point.

Assignment 3 - project status report*, project budget*, existing system USD, DFD, ERD

Note: The PROJECT PRELIMINARY SESSION occurs at this point.

Assignment 4 - team member task file*, project dictionary*, new system USD

Assignment 5 - new system DFD, ERD, GUID

Assignment 6 - new system menu tree, system flowcharts, revised DFD, ERD, hardware, and software specifications

Assignment 7 - system menu and screen form prototypes, cost/benefit analysis, and rationale

Note: The DESIGN REVIEW SESSION occurs at this point.

Assignment 8 - master file maintenance prototype

Note: The PROTOTYPE REVIEW SESSION occurs at this point.

Assignment 9 - training session outline

Assignment 10 - conversion plan summary

Note: The TRAINING SESSION occurs at this point.

Note: The FINAL REPORT is due at this point.

* These assignments require multiple submissions to earn full credit. For example, there will be several "Q&A" documents over the first few weeks of your project. Submit updated project status reports and project budgets EVERY Friday. Your updated team member task files and project dictionaries will be inspected at the time of the midterm and final exams.

Detailed Team Assignments

Team Assignment #1:

1. Decide on a name for your team's consulting firm.

2. Develop a company logo using a graphics package.

3. Integrate this company logo into a company letterhead, which you will use on all correspondence with your client. **Submit** a copy of your stationery.

4. Create an advertising flyer for your consulting company. **Submit** a copy of your flyer.

Team Assignment #2:

1. Read the project packet distributed to your team.

2. **Submit** an initial response to your client (use your letterhead stationery) to include the following:

a. receipt and acceptance of the project
b. restate the client's problem
c. your project milestone timetable
d. your billing policy

3. **Submit** a list of questions regarding the systems, data, procedures, etc. related to your project. These questions should be typed and submitted under a separate cover letter to the client as soon as possible. You will receive a prompt written reply to your first set of questions. It is expected that there will be a series of "Q & A" correspondence, stretching over the first few weeks of the project.

Team Assignment #3:

1. Review the process, data and system modeling discussions in textbook Chapters 4 and 5.

2. **Submit** computerized sketches of the existing system DFD, USD, and ERD.

3. Review the status reporting and budgeting sections of textbook Chapter 6, "Project Management".

4. Develop a draft version (pencil ok) of your budget for the project. Your budget should show estimates, actuals and cumulative over/under amounts, by cost category, on a weekly basis. **Submit** a computerized version of your budget.

6. Develop a draft version (pencil ok) of your project status report for the team project. Your status report should show the following for each activity: start/stop periods, % complete, status summary. **Submit** a computerized version of your project status report.

Note 1: You should submit a revised budget and project status report every Friday until the completion of the project.

Note 2: A budget template (budget.wb1) and a project status template (status.wb1) are on your student resource disk. These files were created in Quattro Pro for Windows.

Team Assignment #4:

1. Review textbook Chapters 6 and 7.

2. **Submit** a computerized sketch of the new system USD.

3. Develop a team member task file to keep track of the different tasks assigned to each team member. You might consider the relevant discussion in textbook Chapter 6 as you design this database file. Enter all of the task information to date. **Submit** a report of this file.

4. Develop a project dictionary to keep track of the different products, files, data elements, etc. associated with your project. Consider the specifications presented in textbook Chapter 6 as you design this database file. **Submit** a report of this file.

Team Assignment #5:

1. Review textbook Chapters 8 and 9.

2. **Submit** a DFD for the new system design.

3. **Submit** an ERD for the new system design.

4. **Submit** a series of GUIDs for the new system design.

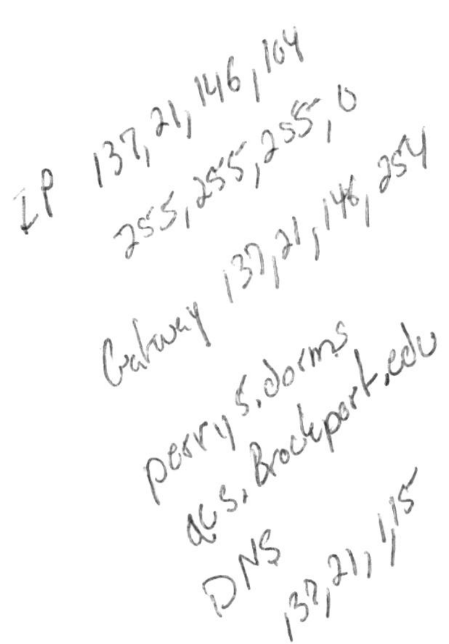

Team Assignment #6:

1. Review textbook Chapter 10.

2. **Submit** the new system menu tree and system flowcharts.

3. **Submit** any revisions to the new system DFD and ERD.

4. Review textbook Chapter 11. **Submit** the hardware and software specifications for your project. This should include as much detail as possible about the products, vendor, prices, shipping costs and taxes.

Team Assignment #7:

1. Review textbook Chapter 11.

2. **Submit** a "Cost/Benefit Analysis" for your project.

3. Develop a prototype of the system menu for your information system. Be prepared to demonstrate this prototype in the computer lab.

4. Develop a prototype of one subsystem's forms or GUIDs. Be prepared to demonstrate this prototype in the computer lab.

5. **Submit** screen dumps and the supporting computer code for these prototype products.

Note 1: There is a cost/benefit template (cost_ben.wb1) on your student resource disk. This file was created in Quattro Pro for Windows.

Team Assignment #8:

1. Review textbook Chapter 12.

2. Develop a prototype of the maintenance processing for one of the master files in your information system. Be prepared to demonstrate this prototype in the CIS lab.

3. **Submit** screen dumps and supporting computer code for these prototype products.

Team Assignment #9:

1. Review textbook Chapter 16.

2. **Submit** an outline of your project training session. This outline should provide details about what you plan to discuss, demonstrate and instruct during your 40 minute session. Identify the team member who is be responsible for each segment of the session.

Team Assignment #10:

1. Review textbook Chapter 17.

2. **Submit** a brief summary of your project conversion plan.

GUIDE TO EFFECTIVE TEAMWORK

The systems analyst rarely works alone. Consultations with managers, users, vendors, and technical support personnel are routine. To some degree, you will experience this type of interaction during your project as you work with teachers and computer lab technicians. For the most part, these dealings are purposeful and productive. However, your work as a member of a project team is never as straight forward. Inevitably, members approach their assignments with differing skills, attitudes, and work habits. This section presents some guidelines to help you and your team work together more effectively.

Team Formation and Leadership

There are many ways to form and organize project teams. Teams can be self-selected, appointed by an outside party (your instructor), or chosen by a team leader. In some cases, teams are formed simply on the basis of member availability. Given that the completion of an information systems project requires many different skills, it is reasonable to assume that team membership is also based on the skills of potential members. Thus, a four member team appointed by your instructor might be composed of one person skilled in database processing, another person with programming experience, a third person well-trained in spreadsheet design and development, and a fourth person talented in word processing and graphic communications. In some respects, the team's organizational structure is obvious in this example - members are assigned duties based on their individual skills. But, who coordinates these efforts? Who assigns tasks that require multiple skills? In short, who provides project leadership?

Team leadership responsibilities can be filled in many ways. A traditional approach is to designate one member to assume all leadership duties throughout the project. Shared leadership is another approach, of which one interesting variation requires each member of the team to fill the leadership role at some time during the project, depending on the activity. For example, the database person might serve as the database administrator, directing other team members to research user needs, develop test files, and write operating instructions. The programmer, on the other hand, might control the detailed processing design activities for all subsystems. And, if the word processing and graphics expert is responsible for all user interface design, he or she might assume the leadership role for all meetings and review sessions between the team and the user. Regardless of the approach, it is important to clearly define the roles and responsibilities of each team member.

Establishing Team Ground Rules

One common complaint about team projects is that some members don't "pull their own weight." This situation is made worse when all members of the team are treated equally when it comes time to pass out rewards or punishments based on the team's performance. Although it is impossible to completely eliminate this problem, there are some remedies that will reduce its severity.

1. **Work to each member's strength.** In line with the assumption that team members possess different skills, it is wise to assign tasks based on those skills. Success can be a narcotic. By experiencing success in one area, team members who might otherwise be inclined to underachieve, may in fact overachieve.

2. **Respect each member's differences.** Often quick to identify a problem and design a remedy, computer professionals are sometimes slow to recognize the value in other, more

methodical or introspective approaches. By the same token, team members who don't necessarily leap to a solution should not stop thinking as soon as one is suggested. The best solutions are usually the result of collaboration. In this way, what may have once seemed to be either overly passive or aggressive behavior is perceived as a valued piece of the process.

3. **Communicate with other team members.** Misunderstandings are often caused by ignorance, especially with respect to individual work habits. Behavior that appears to demonstrate a total lack of concern may be attributable to something completely different. Don't trust non-verbal communications to inform you about how your team members approach a task. Even the simple question, "How are you doing on this task?", can encourage the most recalcitrant team member to respond.

4. **Establish a team protocol.** The operational dynamics of the team project greatly influence your level of success with items 1-3 above. Two important operational concerns involve the dynamics of team meetings and the agreed upon standards of conduct for team members. Although management and social science professionals offer many theories on these topics, we are primarily concerned with some practical guidelines to help your project team work effectively. In that regard, Figure 1-2 might serve as the agenda for your first team meeting. While subsequent meetings might naturally focus on specific project activities, they should also include agenda items that address standards of conduct issues. For example, if your group agrees that team members can and should ask for help when they fall behind schedule, you might add a regular agenda item called "Task Status and Help Request." Alternatively, if your group decides that it will discipline habitually unacceptable behavior, you might include an agenda item on performance review. Perhaps what is most important is that the group should openly evaluate its own performance and, when necessary, initiate action to bring an individual member or the entire team back on course.

Team Project Organizational Meeting
AGENDA

1. Introductions (9 min.)
    a. Computer skills
    b. Personal schedules and phone numbers
2. Selection of a team meeting facilitator (1 min.)
3. Discussion of team project packet (15 min.)
4. Discussion of team protocol (15 min.)
    a. Team meeting day, time, and place
    b. Agenda format for future meetings
    c. Standards of conduct
5. Distribution of initial task assignments (5 min.)
    a. Task description
    b. Deliverable format and content
    c. Responsible team member or members
    d. Due date
6. Adjourn

Figure 1-2

Team Member Profile Sheet

One way to document the different schedules and skills of team members is to assemble a portfolio of team member profiles. This will help the team decide when to schedule meetings, how to distribute project tasks, and how to provide help to members if and when help is needed. Figure 1-3 presents a basic team member profile worksheet.

Team Member Profile

Name ____________________________ Phone ____________

Availability Schedule:

Time	Mon	Tue	Wed	Thr	Fri	Sat	Sun
------	------	-----	------	------	-----	-----	------
------	------	-----	------	------	-----	-----	------
------	------	-----	------	------	-----	-----	------

Computer Related Course Work and Experience:

Introductory	Applications	Programming
literacy	word processing	Pascal
problem solving	spreadsheet	C
operating systems	database	BASIC
networking	graphics	COBOL
----------------	--------------	-------------
----------------	--------------	-------------
----------------	--------------	-------------
----------------	--------------	-------------

Figure 1-3

Team Member Self/Peer Evaluation

Sometimes a formal evaluation instrument is useful to assess the teams performance. Figure 1-4 presents a sample self and peer evaluation worksheet.

Self/Peer Project Team Evaluation

Your Name ____________________ Project ____________

This evaluation is limited to the team project. Complete a section for yourself and each of your team members. Circle the number of points you would award for each performance category.

Self evaluation:

Leadership contribution	1 2 3 4 5
Cooperativeness	1 2 3 4 5
Dependability	1 2 3 4 5
Workload share	1 2 3 4 5
Skills improvement	1 2 3 4 5

Total points ____

Team member ____________________

Leadership contribution	1 2 3 4 5
Cooperativeness	1 2 3 4 5
Dependability	1 2 3 4 5
Workload share	1 2 3 4 5
Skills improvement	1 2 3 4 5

Total points ____

Figure 1-4

DISK FILES

The student resource disk contains several files that will help you complete the team project. Each chapter in the casebook itemizes the files that relate to the material in the chapter. See Appendix A for a complete listing of the files.

Note: Use the following table to correlate file extensions with the product used to create the file:

File Extension	Product
.bmp	Windows Paintbrush
.wb1	Quattro Pro for Windows
.doc	WordPerfect (5.1) for Windows (TT Font: Century Schoolbook)

weeks.bmp This file contains the project milestone events (Figure 1-1).

status.wb1 This is the project status template file.

budget.wb1 This is the project budget template file.

cost_ben.wb1 This is the cost/benefit template file.

agenda.doc This is the team project organization meeting agenda (Figure 1-2).

profile.doc This is the team member profile form (Figure 1-3).

eval.doc This is the self/peer project team evaluation form (Figure 1-4).

CHAPTER 2

THE PROJECT ENVIRONMENT

THE COMPUTER LAB

Most likely, your team will use the school's computer lab to design and develop project deliverables. In some cases, personal computer resources may be available. These resources present an obvious constraint to your project work—you can't design, develop, and implement a computer information system for hardware and software you don't have access to. Thus, one very necessary task is to document the computer resources that are available to your team. The following samples offer a way to begin this task.

Hardware Resources

Figure 2-1 presents a sample worksheet for recording information about the hardware resources available to your project team.

Project Hardware Resources

microprocessor ________	memory cache ________
clock speed ___________	RAM ________________
hard disk space ________	disk cache ____________
video bus _____________	monitor ______________
printer _______________	modem _______________
CD-ROM _____________	scanner ______________
____________________	____________________
____________________	____________________
____________________	____________________
____________________	____________________

Figure 2-1

Software Resources

Figure 2-2 presents a sample worksheet for recording information about the software resources available to your project team.

Project Software Resources

System software specifications:

network operating system ______________________

peer-to-peer _____ client/server _____

local operating system _________________________

desktop manager software ____________________

utility software ______________________________

Application software specifications:

word processing ______________________________

spreadsheet ______________________________

database ______________________________

graphics ______________________________

other ______________________________

Programming software specifications:

procedural ______________________________

nonprocedural ______________________________

Figure 2-2

Operating Procedures

Figure 2-3 presents a sample worksheet for recording information about the operating procedures that govern the hardware and software resources available to your project team.

Project Environment Operating Procedures

Lab access procedures:

network account name __________ password *******

local account name _____________ password *******

team disk area path ____________________________

lab hours: Mon Tue Wed Thr Fri Sat Sun

------ ----- ------ ----- ----- ----- -------

special privileges or restrictions:

____

____

Software access policy:

site licensed software _____________________

student licensed software __________________

Figure 2-3

PROJECT MANAGEMENT

In many ways your team will need to operate like a small enterprise over the course of the next few months. You will need to present yourself to your client through correspondence, project documentation, presentations, and demonstrations. Your instructor, acting as the client, will respond to your questions and provide feedback on your project submissions. In addition, individual team members will develop materials that other members need to access from time to time. The following discussion presents several tools to help you manage the flow of information both internally (among team members) and externally (between the team and client). Of course, if you use project management software, such as Microsoft Project, you won't need to use the sample work sheets, templates and database files.

Project Binder

A project binder contains everything related to the project. Someone on the team should be responsible for maintaining a systematic project document filing system, located in a place accessible to all of the team members. Figure 2-4 presents a sample table of contents for the team project binder.

Project Binder Table of Contents

I. Project Team
1. Team Member Profiles
2. Team Logo and Letterhead

II. Project Environment
1. Hardware and Software Resources
2. Operating Procedures

III. Project Description
1. Project Packet
2. Question and Answer Correspondence
3. Project Contract

IV. Project Management
1. Project Status
2. Team Member Task Assignments
3. Project Disk Space Directory Tree
4. Team Meeting Minutes

V. Team Assignments

VI. Project Deliverables
1. Project Preliminary Session
2. Design Review Session
3. Prototype Session
4. Training Session
5. Final Report and Documentation

Figure 2-4

Project Status

On a regular basis the team will need to monitor its progress on the numerous project tasks. Textbook Chapter 6 presents a detailed discussion of project management, of which the project status report is an integral part. The project status report, however, is a high-level summary the team's progress on the major phases of the project. To keep track of all the small piece parts of the project, you may find it useful to keep task-level records. Figure 2-5 presents a blank project management worksheet, which you can use in a variety of ways. It can serve as a source document for maintaining the project status report spreadsheet. Team members can use it to record the hours they spend on different project activities. It can serve as a detailed team task list. As mentioned in Chapter 1, a project status template (status.wks) is on your student resource disk.

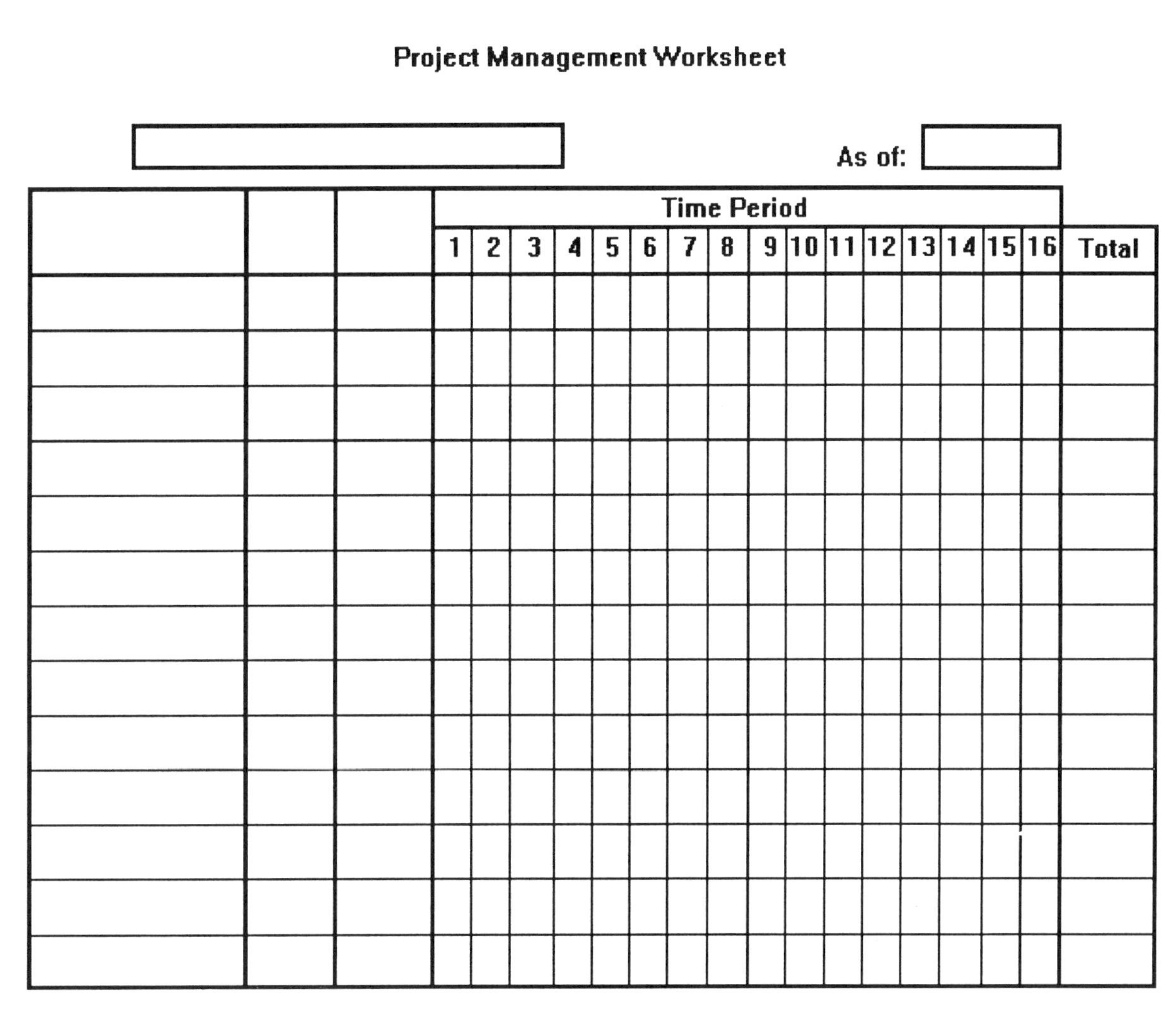

Project Management Worksheet

As of:

			Time Period																
			1	2	3	4	5	6	7	8	9	10	11	12	13	14	15	16	Total

Figure 2-5

Project Task File

The project task file is a detailed account of the responsibilities assigned to team members. The task file not only provides information required to maintain the project status report and the project budget, it also provides a historical record of the project activities. This record can help the team evaluate how well it has distributed project work. Further, it may suggest modifications to the time estimates and member assignments of some remaining tasks. Figure 2-6 provides a sample database file structure to store this information. The file is included on your student resource disk (tasks.dbf). In the interest of simplicity, this file structure does not adhere to strict relational database design rules. It will generate multiple records for a particular task if more than one analyst works on a task or an analyst reports time (hours_wk) to the same task during more than one reporting period (rpt_per).

Task File Structure

Field Name	Type	Size	Decimal	Index
task_id	C	3		Y
task_desc	C	15		N
analyst	C	15		Y
hours_wk	N	6	2	N
rpt_per	C	2		N

Figure 2-6

Project Disk Space

During the project your team will create numerous computer-based products, such as correspondence, diagrams, databases, programs, status reports, budgets, hardware and software specifications, and art work. Team members may need to access some or all of these files at various times during the project. It is important that your team follow a well-defined file naming and disk storage procedure so that everyone will be able to retrieve files when they are needed. Textbook Chapter 6 presents a detailed discussion of a database solution to this problem. Here, we present a simplified alternative, which is to use the hierarchical disk directory scheme used by your computer's operating system. In the case of MS-DOS, the TREE command will display this hierarchy very nicely. Figure 2-7 presents a portion of a sample directory hierarchy for your project. Of course, you can ignore this discussion if you are using a CASE tool that provides this service.

Project Directory Tree

```
C:
____TEAM
       ______PRJMGMT
                    statwk1.wks
                    budwk1.wks
                    tasks.dbf
       ______ARTWORK
                    logo.bmp
                    letterhd.doc
       ______CORRES
                    accept.doc
                    quest1.doc
                    quest2.doc
                    quest3.doc
                    contract.doc
       ______MODELS
                    old_dfd1.bmp
                    old_dfd2.bmp
                    old_usd1.bmp
                    old_erd1.bmp
                    old_mnu1.bmp
       ______DBSUBSYS
                    custmf.dbf
                    main.prg
```

Figure 2-7

PROJECT PACKETS

The project packet briefly introduces a small-enterprise, describes several computer information system needs, and presents a few sample forms used by the enterprise. You will need to supplement these packets with information obtained through the fact-finding activities of the analysis phase of the project. The second team assignment directs you to submit a list of questions about the enterprise and its information needs. This is just the beginning of a discovery process that should continue for several weeks. Industry research and personal experience can provide additional information. In some cases, you may be able to schedule an on-site visit of the actual enterprise.

KNCT Radio

KNCT is a small-market, "easy hit" format FM radio station broadcasting from 6 am to midnight, 7 days a week. They subscribe to music, news, weather and small feature programming services. In addition, there are a few regularly scheduled, locally produced programs. Although disk jockeys preside over the broadcasts most of the time, the station can operate without human intervention.

Commercial spots are mixed with the entertainment programming in typical fashion. From time to time, special event programming may preempt the regular schedule.

The owner/station manager is well versed in computer technology and information systems. At present, the station uses a "286" system for advertiser billing and broadcast scheduling.

The owner has committed $10,000 to upgrade the station's computer information system. The following systems are to be operational by the end of the academic term:

1. Advertiser Master File Maintenance
2. Advertiser Correspondence
3. Revenue Forecasts
4. Broadcast Schedule/Log and Analysis

Advertising revenues (broadcast and production fees) provide 95 percent of the station's income. The other 5 percent comes from special off-air promotions, public appearances, and special one-time audio production services. Advertising rates are based on the length of the ad and the scheduled broadcast time slots. Figure 2-8 presents a sample advertising order form. The time slot notations indicate early, mid, and late two hour time periods during the morning, afternoon, and evening. Thus, EM stands for 6:00 – 8:00 am, MM stands for 8:00 – 10:00 am, LM stands for 10:00 – Noon, and so on. KNCT does not promise to air an advertisement at a specific time, only a specific time slot.

Each 30 seconds of air time costs $12. The weekday early morning (EM), early afternoon (EA), and early evening (EE) time slots cost an extra $3 per 30 seconds. KNCT's in-house production costs are $35 per hour, plus materials (tapes). All music royalties are passed on to the appropriate music publisher or distribution service.

Broadcast Sales Order

Sponsor Name ______________________________ $10,000 Bug- operational before ... of Term

Billing Address ______________________________

Broadcast Description ______________________________

Start Date ____________________ Stop Date ____________________

X = An Extra $3 per 30sec

6-8 8-10 10-12

Time Slot: EM MM LM EA MA LA EE ME LE

X X X

30sec $12

Days: Mon Tue Wed Thr Fri Sat Sun

In House Production $35 hr + materials

Length (air time) ____________

Production Services Required: Copy Voice Music

==

Air Fee Quote ____________ Agency Discount (10%) ____________

Air Notes ______________________________

____

____

Production Fees: Copy _____ Voice _____ Music _____

Production Notes ______________________________

____

Assigned Advertisement Id ______________________________

Figure 2-8

Figure 2-9 presents a sample of a typical program schedule for the station. While there are variations in the schedule from day to day, much of the programming is the same every day. The event number is assigned sequentially in increments of ten so that last minute additions can be penciled in w event numbers. The source field indicates the electronic unit that supplies the broadcast al. For example, Sign-On, Station Id, and Sign-Off all come from tape source unit number one. ro source indicates a live program feed. Source 2.126 indicates that the advertisement for Jack otball is the 126th item on tape source unit number two.

Program Schedule

Date: ____________

Slot/Time	Program Title / Sponsor	Length	Event#	Source
EM 6:00:00	Sign-On	0:00:30	01000	1.1
EM 6:00:30	SatFeed: Weather	0:00:30	01010	3
EM 6:01:00	Music: Spot 1	0:13:00	01020	6
EM 6:14:00	Ad: 126 Jack Football1	0:01:00	01030	2.126
EM 6:15:00	Ad: 45 Goose Cafe	0:00:30	01040	2.45
.				
.				
.				
.				
EA 13:05:50	Station Id	0:00:10	01620	1.6
EA 13:06:00	Live: City Mgr.	0:15:00	01630	0
EA 13:21:00	PSA: Library Book Sale	0:00:30	01640	4.17
EA 13:21:30	Ad: 126 Jack Football	0:01:00	01650	2.126
EA 13:22:30	Ad: 45 Goose Cafe	0:00:30	01660	2.45
EA 13:23:00	Music: Spot 26	0:07:00	01670	6
.				
.				
.				
.				
EE 19:15:00	Live: Jack Football	2:45:00	01950	0
.				
.				
.				
.				
LE 23:30:30	Sign-Off	0:00:30	02450	1.13

Figure 2-9

Although customer billing is not part of the information system request, Figure 2-10 presents a sample advertiser billing worksheet to illustrate the revenue generating elements of the enterprise. Billing adjustments are made when station ratings dip below promised levels. Agency discounts are applied to advertisements contracted through an advertising agency.

Advertiser Billing Worksheet

Sponsor Name ____

Billing Period ____

Advertisement Id ________ Air Spots Scheduled ______________

==

Air Spots Played _____________ Adjustment ________________

Air Fee ________________ Agency Discount ________________

Production Fees: Copy ______ Voice _____ Music _____

Billing Notes ____

____

____

Amount Billed __________ Date __________________________

Figure 2-10

The Reading Association

The Reading Association (TRA) provides events and activities that promote reading to the entire community, as well as to those who make a career of teaching or promoting reading. Local membership, which numbers over 120 members, includes classroom teachers, reading specialists, parents, librarians, professors, journalists, adult literacy volunteers, administrators, and anyone else who places a high value on reading and wishes to promote world-wide literacy.

Membership is on a year-to-year basis. The annual dues are $45, $10 of which is forwarded to the State Reading Association to pay for the member's membership in the State Reading Association and subscription to *READ!*, a bi-monthly magazine.

The officers of TRA want you to develop several products for their computer information system. At present, their only computer application involves a membership database. They use an Apple IIc computer and AppleWorks to keep track of the membership. They have received a $10,000 grant to improve their information system. The system is to be operational by the end of the academic term.

TRA has identified the following information needs:

1. Membership Control and Dues Billing
2. Membership Correspondence
3. Budgeting
4. Event Planning

The major source of revenue, other than special grant funding, is the member dues. Event registration fees, t-shirt sales, and occasional workshop sponsorships provide about 35 percent of the revenues. Expenses include state membership fees, event expenses (facility rental, refreshments, program reproduction, etc.), a $250 scholarship awarded to an outstanding student teacher, state conference expenses for TRA officers, and other miscellaneous items (postage, office supplies, etc.).

The event planner should provide a convenient way to plan and account for the many facets of running an event like "Reading in the Park." Each event is coordinated by a TRA member, who usually delegates responsibility for publicity, set-up, and clean-up to other volunteers. In addition to the date, time, and place of the event, it is important to keep some record of expenses and revenues associated with the event.

Figure 2-11 illustrates the fall flyer distributed to all educators in the county. The association sends out several other mailings during the year. To save postage, the secretary bundles mailings by school site delivery by courier.

The Reading Association's Fall Flyer

The Reading Association cordially invites you to join our organization. Through your membership you will be part of a group of concerned and enthusiastic teachers, librarians and parents. Our goal is to encourage and promote lifelong literacy. You will automatically become a member of the State Reading Association, and receive the bi-monthly magazine READ!

Please plan to begin your new year with us at the Mason Grill at Noon on Saturday, October 15. Virginia Peat will be our invited guest. She will speak on "A New Literacy: Video Games." The luncheon fee is $8.50 per person, payable at the door.

Many other events are planned for the year. Here is a sample.

November 7-9 - State Reading Association Conference.

December 3 - "Read to Me" with Chris Peters.

January 25 - "Literacy By The Fireside" with Ron James.

February 24 - "The Authors' Festival".

April 21 - "Reading in the Park"

May 5 - Reading Association Banquet.

To join the Reading Association , please complete the attached form and mail it to Paula Grimes, Treasurer, 1212 Hector, Eureka, CA 95501. Dues for 1995/96 are $45.00.

We look forward to seeing you Saturday!

Figure 2-11

The membership form is illustrated in Figure 2-12. In addition to a brief interest and experience inventory, the form includes information regarding the member's affiliation (school, business, parent, etc.), membership status (renewal, new), and State Reading Association number (for renewals only). Membership terms run from August to July to conform with the school year.

The Reading Association Membership Form

Name ____

Home Address / Phone ____________________________

Work Address / Phone ____________________________

Affiliation ___________________________________

Renewal Membership __________ New Membership ________

State Reading Association # ________

I am interested in helping with the following events ____

____

____

I have experience in the following areas:

Publicity ___ Grant Writing ___ Public Speaking ____
Computer ___ Other ____________________________

Please make your check for $45.00 payable to "The Reading Association" and note that it is for 1995/96 dues.

Figure 2-12

Figure 2-13 presents an event worksheet. Event date, time, location, and coordinator is posted on a master calendar in the TRA office.

TRA Event Worksheet

Event ______________________ Date(s) ___________

Time(s) ___________

Location ___________

Description ____________________________________

____

____

Coordinator ___________________ Phone ___________

Helpers ______________________ Phone ___________

___________________ Phone ___________

___________________ Phone ___________

Budget:

Cost Description	Amount	Revenue Description	Amount
_____________	_____	_______________	_____
_____________	_____	_______________	_____
_____________	_____	_______________	_____
_____________	_____	_______________	_____
_____________	_____	_______________	_____
Total	________	Total	________

Figure 2-13

Associated Students, Inc.

Associated Students, Inc. (ASI) is the umbrella organization for student government, as well as numerous student clubs, charities, and profit making entities. ASI wants to develop a computer-based student club information system, which it will offer to student clubs at its home campus at no charge. If this is successful, ASI plans to market the system to student organizations around the state.

Student clubs sponsor many activities throughout the school year (eg. fund raising, guest speakers, awards banquets, scholarships, field trips, etc.). The typical club carries an active membership of 100 to 150 students, from which club officers (president, vice-president, secretary, treasurer) are elected.

While the officers interface with other club officers and ASI, most clubs use a subcommittee structure to manage their internal affairs. The membership committee handles all membership record keeping, which mostly involves membership lists and correspondence. The financial management committee is responsible for the receipt and dispersal of monies, as well as fund raising activities. The presentations committee arranges for guest speakers, field trips, and other special events. The scholarship committee is in charge of awards and scholarship events. Each committee is populated with four or five members, with one person appointed as chair.

ASI has budgeted $10,000 for the computer information system. It plans to locate the computer in the Associated Students office on campus. Student clubs can use the computer hands-on or via a modem to access their respective systems. ASI would like you to develop the following products by the end of the term:

1. Membership Correspondence
2. Membership and Committee Databases
3. Scholar Tracking System
4. Revenue/Expense Budgets and Forecasts

Figure 12-14 presents a club/student activities data sheet used by several clubs. This form provides information about club members' dues and committee history. Members are supposed to fill out the top portion of the form whenever their address information changes.

Club ____________
Membership Form

Member ______________/____/______________________
First MI Last

Address Information: New ___ Change ___ Date ________

Campus ________________/___________/___________
Street City Zip

Home ________________/___________/___/_______
Street City State Zip

Phone
Campus _____________ Home (____)_____________

==

Dues Paid: Year Amount Date Method

________/_______/_______/________

________/_______/_______/________

________/_______/_______/________

________/_______/_______/________

Committee Work:

Committee Date Activity Description

________________/_________/___________________________

________________/_________/___________________________

________________/_________/___________________________

________________/_________/___________________________

Figure 2-14

The scholar tracking system is a new feature of ASI's information gathering and reporting system. Usually each club sponsors three sophomore scholars to attend a three week inter-session at another college. The tracking system will contain a initial profile of each participant, along with annual follow-up information for three years. Figure 2-15 presents a preliminary sketch of the data elements included in the profile and follow-ups.

Scholar Tracking Data Sheet

Scholar Name ______________/____/__________________
First MI Last

Inter-session Dates _______________/______________
Start Stop

Club Sponsor ______________________________________

Scholar Profile:

Age _____ Field of Interest ____________________

Units Attempted __________

Units Completed __________ Cummulative GPA _______

Follow-Up:	Year 1	Year 2	Year 3
Units Attempted	_____	_____	_____
Units Completed	_____	_____	_____
Cummulative GPA	_____	_____	_____

Figure 2-15

The Repertory Theatre

The Repertory Theatre (TRT) is in its 20th season of presenting a wide variety of quality, live theatre to the residents of a small suburban community.

They present five regular productions a season, plus one summer production in conjunction with the local community college. Each production runs for about three weeks, with five performances per week, plus a special opening night performance. Overall, occupancy is about 75 percent of capacity, with the typical audience composition 55 percent subscribed and 20 percent walk-up.

The lead and major supporting actors are paid a nominal amount for each performance, as is the regular staff at the theatre. Each production is budgeted for permissions, sets, costumes, lighting, sound, stage managment, direction, and publicity. Ticket revenues cover about 70 percent of the expenses. Interest from an modest endowment, local fund raising events, and member donations account for the remaining expense amounts. The theatre usually has a small surplus at the end of the season.

Recently, a prominent resident bequeathed $10,000 to the theatre with the stipulation that the money be used to improve communications with the community, increase season subscriptions, and stream-line ticket handling. The Board of Directors would like you to develop the following products by the end of the term:

1. Subscriber and Member Correspondence

2. Subscriber and Member Databases

3. Ticket Management

4. Occupancy Analysis

Figure 2-16 presents a subscription order form.

The Repertory Theatre
Season Subscription Form

Name ____

Address ____________________ City ____________________

State _____ Zip __________ Phone ______________________

Subscription Preference:

Subscription Plan	Ticket Cost
A: Opening Nights	____ X $80 = _____
B: Friday or Saturday Night Series	____ X $55 = _____
C: Thursday Night Series	____ X $40 = _____
D: Saturday or Sunday Matinees	____ X $40 = _____
E: Special Preview Subscription	____ X $30 = _____
	Total Cost =_______

Seating Preference:

Choose One: Front to Middle _____ Middle to Back _____

Choose One: Left _____ Center _____ Right _____

Special Memberships:

Patron ($50-$99) _____ Donor ($250-$499) _____

Sponsor ($100-$249) _____ Benefactor ($500+) _____

Billing:

Check Enclosed ______

Charge to Visa: Card Number ___________ Expiration _____

Figure 2-16

Figure 2-17 presents the seating arrangement and numbering system.

Seating Arrangement

Stage

	Left	Center	Right
Row A:	8 6 4 2	12 14 16 18 19 17 15 13 11	1 3 5 7
Row B:	8 6 4 2	12 14 16 18 19 17 15 13 11	1 3 5 7
Row C:	8 6 4 2	12 14 16 18 19 17 15 13 11	1 3 5 7
Row D:	8 6 4 2	12 14 16 18 19 17 15 13 11	1 3 5 7
Row E:	8 6 4 2	12 14 16 18 19 17 15 13 11	1 3 5 7
Row F:	6 4 2	12 14 16 18 19 17 15 13 11	1 3 5
Row G:	8 6 4 2	12 14 16 18 19 17 15 13 11	1 3 5 7
Row H:	8 6 4 2	12 14 16 18 19 17 15 13 11	1 3 5 7
Row I:	8 6 4 2	12 14 16 18 19 17 15 13 11	1 3 5 7
Row J:	6 4 2	12 14 16 18 19 17 15 13 11	1 3 5
Row K:	8 6 4 2	12 14 16 18 19 17 15 13 11	1 3 5 7
Row L:	8 6 4 2	12 14 16 18 19 17 15 13 11	1 3 5 7
Row M:	8 6 4 2	12 14 16 18 19 17 15 13 11	1 3 5 7
Row N:	8 6 4 2	12 14 16 18 19 17 15 13 11	1 3 5 7
Row O:	8 6 4 2	12 14 16 18 19 17 15 13 11	1 3 5 7
Row P:	8 6 4 2	12 14 16 18 19 17 15 13 11	1 3 5 7
Row Q:	6 4 2	12 14 16 18 19 17 15 13 11	1 3 5

Figure 2-17

DISK FILES

A new file format is added to and used in this collection of student resource disk file—the database file, with a .dbf file extension.

hardware.doc This is the project hardware resources worksheet (Figure 2-1).

software.doc This is the project software resources worksheet (Figure 2-1).

operate.doc This is the project environment operating procedures worksheet (Figure 2-3).

gantt.bmp This is the project management worksheet bit mapped file (Figure 2-5).

tasks.dbf This is an empty dBASE IV file for recording team member task assignments.

PART II

COMPLETING THE TEAM PROJECT STEP BY STEP

This part of the casebook parallels the textbook in its treatment of the small-enterprise computer information system project. In other words, Chapters 3-18 of the casebook provide details and insights into the Cornucopia project, which appears at the end of Chapters 3-18 in the textbook. This material will help you complete your own team project.

Each chapter begins with a brief overview of the major instructional points presented in the corresponding textbook chapter. This is followed by an analysis of the Cornucopia project as it progresses from problem identification and definition to information system maintenance and review. The chapters conclude with extended exercises relating to the textbook's inter-chapter examples.

CHAPTER 3

PROBLEM IDENTIFICATION AND DEFINITION

OVERVIEW

Every computer information system project begins with a general perception that there is a problem with the present computer-based or manual system. We use the word *problem* in the broadest sense—the analyst is challenged to find better ways to satisfy user demands for information products. The first step in meeting this challenge is to clearly identify and define the problem.

It is incorrect to assume that the user will be able to pinpoint the problem or that the problem is confined to a specific individual or computer product. In fact, there may be several interrelated problems that cross over all five elements of a computer information system. Further, the nature of the enterprise may significantly influence the way the problem presents itself.

This leads the analyst to pursue many avenues of investigation before attempting to identify and define the problem. Underlying each approach is the understanding that information must meet the following basic requirements:

Basic Information Processing Requirements

1. Information must be relevant.
2. Information must be accurate.
3. Information must be timely.
4. Information must be useful.
5. Information must be affordable.
6. Information must be adaptable.

After the analyst generally understands the problem, as well as the time and money constraints under which the problem must be solved, it is time to evaluate whether it is practical to even begin the project. This is called a feasibility study.

Even when the feasibility study indicates that the project is likely to succeed, no one may fully understand the details of the proposed project. Nevertheless, the analyst and user must attempt to document the following items in a project contract:

Elements of the Project Contract

1. Problem Summary
2. Scope
3. Constraints
4. Objectives

CORNUCOPIA TO DATE

Cornucopia was chosen as the model small enterprise because the major information products are commonplace. Hopefully you will be able to identify similar product needs in your team project and adapt the Cornucopia solutions to meet those needs.

The Inside Story

Margaret Height's initial letter informs the analysts about the general information needs of the enterprise. There are four: customer record keeping, CD reordering, customer communications, and sales trending. These items are understandable when looked upon individually, but when taken together, they pose a new problem. Just exactly how will an integrated information system satisfy each need without costly and inefficient duplication of data and computer processing? Further, how can an inexperienced analyst estimate the time and money required to create the system or the benefits that such a system will produce? No doubt, your team may face the same dilemma when confronted with the task of creating a contract for your project.

The Cornucopia project analysts make three very critical decisions right away. First, they decide that the first three information products are routine, requiring very straight forward database and word processing solutions. The only unusual information product involves sales trending, which probably will require a spreadsheet solution of some kind. Second, knowing that powerful microcomputer hardware and software is available at prices well below the budget constraint of $10,000, they decide that there will be plenty of money left over to pay their fee, even though they don't know exactly how many hours it will take them to finish the product. Third, they decide to fashion several measurable project objectives by assuming that each information product will either improve sales or reduce enterprise labor hours. All three of these decisions are highly speculative, but they serve to advance the project, as well as set a baseline from which to measure success and failure, which will ultimately enhance their ability to make better decisions in the future.

How to Copy the Concepts

If your team is assigned one of the projects in Chapter 2, the information products are already identified. Problem identification and definition can proceed with a series of "question and answer" correspondence between the team and the instructor. For example, if the need is to "maintain customer records", you need to find out exactly what customer information is relevant to the project. If you are asked to provide a forecast, you need to find out how the forecasted data is presently derived and what data may influence future derivations.

At this point, your primary goal is to understand the project well enough to write the contract. The Cornucopia project contract can serve as the model for this important task. Following the decision making process used by the Cornucopia analysts, your team can focus on two aspects of the contract (scope and objectives). The other elements of the contract (problem summary and constraints) are already fairly well defined.

The question of project scope is particularly important. You don't want to overlook anything or inadvertently add something to the project requirements. Make sure to question just how far each information product extends into other enterprise information systems. For example, if you are required to "keep track of" membership dues, does this mean that you need to include an accounting system to record cash receipts or can you simply add a field to the membership records?

In dealing with project objectives, remember to evaluate each information product in terms of its potential influence on revenue or service enhancement and enterprise labor hours. The more specific you are at this point, the easier it will be for you to develop the cost/benefit chart discussed in Chapter 11. Don't be intimidated by the fact that you really don't know if you can meet these objectives. In order to learn how to develop such estimates, you need some practice, with the understanding that failing in this setting is acceptable, and maybe even expected.

DISK FILES

The student resource disk contains numerous files related to the Cornucopia project. These files are described chapter by chapter, as they relate to the casebook and textbook. Refer to Appendix A for a complete listing of all the files on your disk.

contract.doc This is the Cornucopia project contract that appears in the textbook.

EXTENDED EXERCISES

The extended exercises relate to the inter-chapter examples in the textbook. Each of the last four sections of the text (analysis, design, development, and implementation) contain different examples, as noted below.

Inter-Chapter Examples

Section 2 - Analysis (Chapters 3-6)

* CIS Lab - The Time Keeper System (TKSystem)

* Silhouette Sea Charters

Section 3 - Design (Chapters 7-11)

* Sunrise Systems

Section 4 - Development (Chapters 12-15)

* Political Research Corporation (PRC)

Section 5 - Implementation (Chapters 16-18)

* *LandscapeS*

CIS Lab Extended Exercise 1

a. Review the Memo for File describing the CIS Lab problem (see textbook Figure 3-4).

b. Speculate on which basic information system requirements are not being accommodated by the present card system.

c. Is there anything about these problems that is peculiar to a college or small enterprise environment?

d. Research and describe the existing student usage information system in your college computer lab.

Silhouette Extended Exercise 1

a. Review the Memo for File describing Silhouette Sea Charters' problem (see textbook Figure 3-4).

b. Prepare a Request for Services for this problem.

CHAPTER 4

DATA FLOW DIAGRAMS

OVERVIEW

A data flow diagram (DFD) is a picture of the information system processes and their inputs and outputs, as well as the input sources and output destinations. It is called a process model because it focuses on the transformation of data into information. There are four elements to a data flow diagram:

Elements of a DFD

1. Process
2. Data Flows
3. Data Stores
4. External Entity

The construction of the DFD is the analyst's first attempt to systematically define an information system. To do this, the analyst must conduct an investigation of the current system and then design the new system. The preparation of the DFD for an existing system requires the analyst to interact primarily with the people, procedures and data elements of the system. On the other hand, the DFD for a new system is a product of the analyst's experience, mixed with generous amounts of end user participation. There are two established techniques for developing data flow diagrams:

Creating a DFD

1. Bottom-up Method ...
   using a series of task IPO charts

2. Top-down Method ...
   asking "Then what happens?"

This model building activity illustrates a recurrent theme in systems work: Difficult tasks are solved in small segments. With process modeling, we decompose the broadest picture of the system into succeedingly more detailed pictures of the system. The following progression of diagrams are prepared:

Process Model Diagram Decomposition

1. Context Diagram
2. First-level DFD
3. Second-level DFD
4. ... and so on

Throughout this activity, the analyst is challenged to develop a series of abstractions of the real world. A good abstraction presents a clear picture or description of the essence of an object or system. This requires the analyst to (1) discover what is most important about a subject and (2) find a way to express what may be very complex concepts in a simple, understandable fashion.

CORNUCOPIA TO DATE

After the details of the project contract are worked out, the analyst must develop a clear understanding of the current system. The collection of sample documents, interview and observation notes and industry research findings provide most of the information necessary to construct the context diagram and DFDs for the current system. The Cornucopia project team is now engaged in that process.

The Inside Story

By design, this project is limited in its scope and complexity. This will allow us to focus our attention on the SDLC process, rather than on highly technical and stylized computer systems. However, even with this restriction, the team has difficulty developing the existing system process models.

To begin with, the team becomes confused about the role the owner plays in the existing system. Upon learning that customers often give their mailing list requests to the owner and that the sales clerks report stock shortages to the owner, the team mistakenly classifies the owner as an external entity because she receives information. This decision is reversed once the team realizes that, in this case, the owner merely facilitates update operations to the customer and order data stores.

After the owner stipulates that the sales operation is not part of the project, the team dismisses any thought of including sales in the processing models. When the owner informs the team that formal inventories are done infrequently, they are confused about where ordering information comes from. Only then do they realize that the sales system is involved in the ordering process in a big way, but as an external entity, not as a process within the system under study.

Finally, the most difficult problem concerns the actual sales transaction activity. The team understands that the customer receives inventory items in exchange for money. This reduces inventory, which is replenished when shipments are received from suppliers, who are themselves responding to orders generated by past sales transactions, and so on. How does all of this input and output affect the inventory process? After all, inventory procedures are never mentioned in the original letter, the feasibility report or the project contract. This problem persists until the team clarifies an important point with the owner. Although formal inventories are done infrequently, the continual observations of the employees serves as a process which can be correctly identified as a process within the system.

You may experience similar frustrations as you attempt to diagram information systems of this same simplicity. You should be encouraged by the fact that this process will become easier as you gain experience.

How to Copy the Concepts

Your original set of project documents, the subsequent question and answer correspondence and the project contract should provide the information necessary to develop the context diagram and the first level DFD of your project's existing system. But, you will only discover if this is true when you try to create these models.

If you have great difficulty with this exercise, you must decide if the problem is with your understanding of the current system or if the problem lies with your understanding of the abstraction process itself. Questions about the current system can be answered by resuming the question and answer procedures. Your questions should be as specific as possible. "I don't understand the current system" is not a specific question! If you can't distinguish one process from another, return to the "top-down" approach discussed in the text by asking, "What do you do with this piece of data or this document?" Make a note of the first action word or phrase in the reply, for that is most likely a good name for the first process. Then, you might ask "Where does this data or document come from?" The answer to this question is either an external entity, a data store or another process. Continue this sequence until you can develop a preliminary sketch of the models, after which you can solicit the user's confirmation that indeed you do understand the current system.

Understanding the abstraction process is admittedly a more difficult task. You might try to build your appreciation for this concept by recalling some of the most obvious abstractions that exist in your everyday life. For example, when you are asked to describe your method for preparing a meal, you might reply, "Which meal and for how many people?" Your first impulse is to add some of the specificity back into the original question. In this case, the phrase "preparing a meal" describes an abstract concept. Most of us have no difficulty in telling someone that we are going home to make dinner. We are confident that this abstraction will communicate to the listener, who does not need to know all of the details about the particular dinner we have in mind. If this example helps, then simply reverse the process. Given a set of detailed narrations about how a system works, try to sift the details out, leaving only a broad, generalized statement, or abstraction, of the activity.

On a more practical level, the actual drawing of the diagrams is best performed with a CASE tool. If this product is not available, you have two choices. Either you can do this by hand or you can use a simple paintbrush program. The hand method should be your last resort. Your product will not look professional. The paintbrush method will solve this problem, but create another. Paint programs are restricted to a fairly small work screen. Therefore, you will need to use a paper sketch to plan the layout of your diagram. But, do not waste time trying to make your sketch look pretty, for it is disposable as soon as your paint image begins to take shape.

Another difficulty with the paint program solution is that you do not have an electronic data dictionary to remind you of the labels you have assigned to all of the project elements. This means that you will need to maintain some notation system for these values.

Finally, the paint programs do not usually include image libraries that include the standard symbols for the models you need to develop. To avoid inconsistencies and duplicate effort, you should create your own set of symbols and save them in a file available to all of the project team members.

DISK FILES

To make your work easier, the student resource disk contains several bit mapped image files that are particularly relevant to the process modeling task. These files were created with Windows 3.1 Paintbrush. Included is an image file with all of the DFD symbols used in the text, as well as skeletons of Cornucopia's context diagram and first-level DFD. The skeleton files may help you create similar diagrams for your team project.

dfd_sym.bmp — This file contains images used to render the data flow diagrams in the textbook.

context.bmp — This skeleton image of Cornucopia's existing system context diagram contains three external entities, each with two data flow lines.

dfd_lev1.bmp — This skeleton image of Cornucopia's first-level DFD contains four processes, three external entities, four data stores, and several connecting data flow lines.

EXTENDED EXERCISES

These exercises are extensions of the material presented in the text and the extended exercises that appear in the previous chapter of this casebook.

CIS Lab Extended Exercise 2

Using the CIS Lab example that appears in the text as a guide, investigate the student time reporting procedures in your school's computer lab. Prepare the following:

a. A narrative describing the existing system.

b. A context diagram for the existing system.

c. A series of Task IPO charts that identify the inputs and outputs for each process within the existing system. Be sure to include the sources, sinks, and data stores associated with each data flow.

d. The first level DFD for the existing system.

Silhouette Extended Exercise 2

Silhouette Sea Charter's original memo (Figure 3-4) mentions three areas of concern: scheduling, billing, and payroll. The descriptive narrative for the existing scheduling system appears in Chapter 4 of the text, along with a rough idea of the task IPO chart.

a. Develop the context diagram for the existing scheduling system.

b. Complete the Task IPO chart for the existing scheduling process. Include the sources, sinks, and data stores associated with each input and output.

c. Develop the first level DFD for the existing scheduling system.

CHAPTER 5

SYSTEM AND DATA MODELS

OVERVIEW

This chapter continues the discussion of the modeling process by presenting three system models and one data model. Coupled with the data flow diagram, the analyst prepares an elaborate sequence of diagrams to explain the existing information system and propose a new system. Each model represents a different view, as summarized below:

Differing CIS Model Views

Data Flow Diagram (DFD)	the process view
User's System Diagram (USD)	the user's view
Menu Tree	the operator's view
System Flowchart	the programmer's view
Entity-Relationship Diagram (ERD)	the data view

The user's system diagram (USD) is described as a user-friendly substitute for the data flow diagram. It simplifies the data flow lines and employs familiar icons to help the reader reassociate elements in the abstract model with those in the real world.

The menu tree is a hierarchical display of the operational choices available to the operator. Although these choices usually appear in the DFD and USD as processes, the menu tree illustrates how to navigate from one to the other. This, in turn, identifies the functional modules of the system, which helps the analyst prepare for future detailed system design.

The system flowchart shows the relationships between the system data files and processing software. The analyst uses this model to isolate those portions of the system that we traditionally associate with computer applications. Further, the model shows the file sharing relationships which will help the analyst integrate the applications into a computer information system.

The entity-relationship diagram (ERD) is used to model the file relationships in the system. As such, it is called a data model. File relationships can be described in terms of the data they hold in common. For example, your bank maintains a master file of its depositors. This file contains a unique identification number for each account, along with information about the people authorized to access the account. In addition, the bank maintains a transaction file to record each and every deposit or withdrawal. In order to associate the transactions with the appropriate account, the master and transaction files must have one common field - the account identification number. Thus, even if your bank maintains 10,000 accounts and processes 500,000 transactions a month, it can quickly and accurately produce your account statement by indexing and relating the two data files on the account number. This example describes the classic one-to-many relationship between master and transaction files. The ERD documents file relationships in an easy-to-read two-dimensional format. You must understand the file relationships in your information system if you are to succeed in the upcoming design phase of the project.

To recap, the analyst creates three types of models to describe a computer information system: process, system, and data. Overall, we have studied five specific models, as summarized below:

Types of CIS Models

Process	... data flow diagram (DFD)
System	... user's system diagram (USD) ... menu tree ... system flowchart
Data	... entity-relationship diagram (ERD)

As the analysis process moves from one model to another, you should notice the repetition of the data element descriptions. In the banking example, the account identification number appears in both the master and transaction files. Further, the files should appear in the DFD and USD as data stores, in the system flowchart as data files, and in the ERD as data entities. In all cases, it is important to use the same label, or data element name, to refer to the same object, wherever it may appear in the various models. A special tool, called a data dictionary, is used to document the project's numerous data element names.

CORNUCOPIA TO DATE

The analysts experienced considerable frustration in developing the data flow diagram of the existing system. In fact, they are not completely sure that they understand how the existing system works. This is the time to reconnect with the owner. Unfortunately, they expect that the data flow diagram will be confusing to the owner, so they create a more user-friendly version of the DFD (the user's system diagram) to use in their discussions.

The Inside Story

The meeting between the analyst and owner to review the USD goes well. The owner judges the analysts' rendition of the existing system as accurate, although she is unsure about why the system, which she views holistically, is chopped up into pictures, lines, and labels.

Feeling flush with pride because they have gotten off to a great start, the analysts continue the modeling process by developing the entity-relationship diagram. Three problems arise immediately to dampen their spirits.

First, they notice the DFD shows no direct relationship between supplier and inventory. Remember, the analysts are trying to document the file relationships with the ERD. They had assumed that these two files were related because "inventory items come from suppliers." After some thought, they realize that the order file is the intermediary between the files.

The second problem concerns their choice of data element names. The use of the same label (supplier) to identify a data store and an external entity is confusing. But, since they cannot think of a better alternative, they agree to leave the name as is.

The third problem is that the customer master file does not interact with any other file in the system. Although this looks funny on the ERD, they leave it alone.

Of course, these problems are temporary, which should encourage you when your team runs into similar circumstances. Each activity in the *enhanced* SDLC is capable of producing problems that frustrate your attempts to move quickly from phase to phase. In this case, the major challenge is to normalize the order-inventory file relationship uncovered by solving the problems discussed above.

How to Copy the Concepts

Experience shows that the modeling process presented in this and the previous chapter is the most frustrating part of the team project. This is certainly understandable when you consider that you are using a new concept to describe a system you do not fully understand.

You can reduce this frustration by concentrating your initial effort on the easiest files to identify -- the master files. Look for the objects in the system that "act" on other objects. For example, in the banking problem described above, customers initiate action by opening an account, making deposits, requesting account balances, etc. Thus, account-holder or customer is a good master file candidate. The next step is to define all of the information you think is important to describe an individual customer, remembering to include one field that distinguishes each customer from the others. After this, you should prepare five sample master file records. Now you are ready to move on to the more challenging transaction file.

Transaction files describe an event. A transaction file record might include data describing the time, place, and nature of the event. It usually includes data describing the object that initiated the event, with the object being more fully described by a master file record. If you ask the question "What actions do the objects described by a master file take?", you are likely to find that the answer points to an event. The deposits made by customers in the banking example can be defined as records within a transaction file. After defining all of the information you think is important to describe this event, you should prepare five sample transaction file records. Now, you can define the relationship between these two files.

File relationships can be defined by asking the question "For each record in file A, how many corresponding records may or must exist in file B?". Applying this to the banking example we can say: For each customer master file record there may be one or more corresponding records in the deposits transaction file. Conversely, for each record in the deposits transaction file there must be one and only one corresponding record in the customer master file. We describe such a relationship as one-to-many.

DISK FILES

sys_sym.bmp — This file contains images used to render the system flowcharts in the textbook.

erd_sym.bmp — This file contains images used to render the entity-relationship diagrams in the textbook.

data_dct.doc — This is the data dictionary entry form.

old_erd.bmp — This skeleton image of Cornucopia's existing entity-relationship diagram contains four entities.

EXTENDED EXERCISES

CIS Lab Extended Exercise 3

Refer to the process and system models to identify the data stores and data files in the TKSystem.

a. Which files would you classify as master files?

b. Which of these files would you classify as transaction files?

c. Describe the relationship between these files?

d. Normalize these file relationships, adding new file if necessary.

e. Prepare the entity-relationship diagram for the TKSystem.

Silhouette Extended Exercise 3

Review the narrative and the detailed file layouts for the existing Silhouette Sea Charter information system.

a. Create a user's system diagram.

b. Create a system flowchart.

CHAPTER 6

PROJECT MANAGEMENT

OVERVIEW

One of the analyst's most important responsibilities involves project management, which includes task planning, resource estimation and allocation, and activity monitoring. This chapter presents three important tools to help analysts do this: Gantt charts, PERT charts, and project dictionaries. From the basic Gantt chart we fashion project status and budget worksheets, which help analysts monitor their progress towards the various project goals and subgoals. The PERT chart shows the completion dependencies for the numerous project tasks. The project dictionary catalogs all of the papers, files, and products associated with the project. The following summarizes how these tools can be used.

Using Project Management Tools

Status Report	Helps the analyst monitor project activity completion and compare actual labor hours expended to estimates.
Budget	Helps the analyst monitor project expenditures, as well as improve cost estimating skills.
PERT Chart	Helps the analyst reallocate excess resources to activities that can be completed independently or to behind-schedule activities that endanger the final project completion date.
Project Dictionary	Helps the analyst quickly locate the most recent version of everything from the project contract to the status report.

Estimating the resources required to complete a project presents another challenge to the analyst. Basically, the analyst needs to itemize the hardware and software resources, as well as all of the tasks required to complete the project, without knowing what pitfalls await the project team. Two methods, detailed below, are offered to help the analyst accomplish this.

Estimating Project Costs

Bottom-up Approach	The sum-of-the-tasks method requires detailed knowledge of the time needed to complete individual tasks. The time-and-materials method is usually too open-ended to suit the user.
Top-down Approach	Based on some historical experience, the project budget constraint is divided into labor, hardware, and software cost categories.

This chapter, which completes the "Analysis" section of the project, concludes with a discussion of the "Project Preliminary Presentation." This session, occurring after the Project Contract is in place, informs the user about the analyst's understanding of the enterprise and its information system requirements, as well as the broad design concepts, budget, and completion timetable of the proposed system.

CORNUCOPIA TO DATE

With the introduction of the project status and budget worksheets, the analysts initiate a regular report on their progress towards completing the project. In addition, they periodically compare their estimates of resource requirements with actual expenditures. Figure 6-1 summarizes the budget and status reports as of the third week of the project schedule.

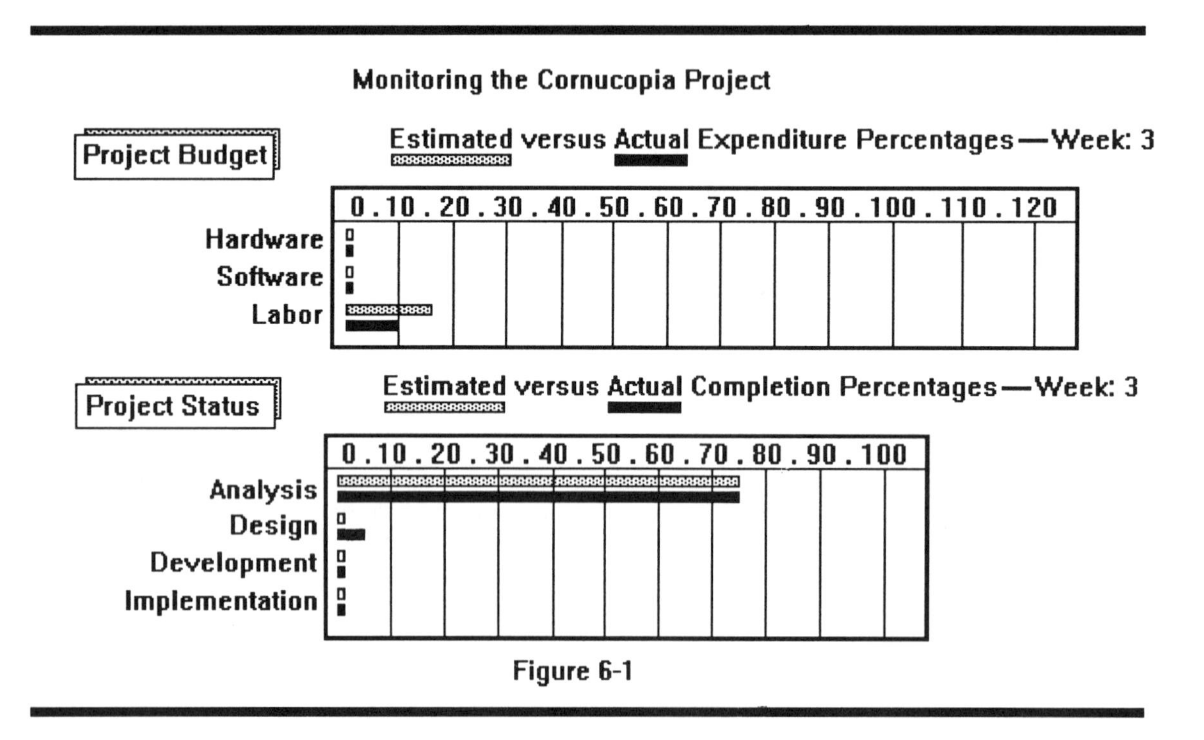

Figure 6-1

The Inside Story

Two project constraints, the $10,000 project budget and 16 week project completion time frame, provide the analysts with very clear project management reference points. They use the top-down cost estimating approach and later switch to a very generalized bottom-up approach to estimate the breakdown of labor hours.

After reducing the cost categories to hardware, software, and labor, the analysts rely on their microcomputer experience to estimate hardware costs at $4,000 and software costs at $1,500, leaving $4,500 for labor costs. Because labor charges are set at $50, they can allocate 90 hours to the analysis, design, development, and implementation phases of the project. The initial allocation is as follows:

Initial Labor Hour Allocation

Analysis	20%
Design	30%
Development	30%
Implementation	20%

The hours are spread evenly over four or five week periods, allowing for a one-week overlap for transition from one phase to another. The detailed task list, which was prepared independently of this budgeting process, is now used to more specifically evaluate labor hour requirements and make the allocation estimates. Figure 6-2 summarizes the allocations.

Labor Hour Allocation by Activity

Analysis (20%)	
Initial consultation	4
Full analysis	14
Design and Development (60%)	
Initial design sketch	7
Create prototype	27
Final product development	20
Implementation (15%)	
Develop system documentation	8
Develop training materials	4
Installation	2

Figure 6-2

This breakdown serves as the basis for constructing the PERT chart that appears as textbook Figure 6-13. Although such a chart is really not necessary for a project of this size, it is instructive to observe the dependent nature of many of the activity sequences.

How to Copy the Concepts

Your project contract specifies budget and time constraints. If these are not the same as those specified for the Cornucopia project, calculate a budget and time adjustment factor by dividing your project constraints by Cornucopia's. Use this factor to reduce or inflate the Cornucopia estimates as a starting point. For example, if your budget is $12,000 and your time constraint is 12 weeks, your adjustment factors are 1.2 (12,000/10,000) for your budget estimates and .75 (12/16) for your time estimates. Thus, you should begin with the following cost and hour estimates:

Adjusted Cost and Hour Estimates

Costs		
	Hardware	$4,800
	Software	$1,800
	Labor	$5,400
Hours		
	Analysis	13.5 hours
	Design and Development	40.5 hours
	Implementation	10.5 hours

You should note that under these constraints, the labor rate is inflated to over $80 per hour. This illustrates that you need to apply your own judgement to the estimates. For example, given the $5,400 labor cost budget and the $50 per hour labor rate, you could allocate 108 hours over the 12 week period in the 20:60:20 ratio as follows:

Further Adjusted Hour Estimates

Hours		
	Analysis	21.6 hours
	Design and Development	64.8 hours
	Implementation	21.6 hours

Even if your project constraints are the same as those for the sample project, you should consider adjusting the Cornucopia numbers to account for special circumstances that apply to your project. To help you improve your estimating skills, it is important to document the rationale used to make these adjustments. This will allow you to conduct a more informed review of your actuals and estimates later on.

It should be apparent that each team member will need to keep accurate records of time spent on each assigned task. Consider using a paper-based system, such as the Billable Hours Log illustrated in textbook Figure 6-3, or a computer-based system, such as the one suggested in casebook Figure 2-6.

Finally, you should develop a consistent strategy for managing numerous project documents, files, software, and so on. Probably the easiest method is to use the operating system's file manager utility as illustrated in casebook Figure 2-7. Alternatively, you can use the database included on your resource disk (proj_dct.dbf).

DISK FILES

Several files, mentioned in earlier chapters, are relevant to your project management activities. They are the Gantt chart (gantt.bmp), the project status worksheet (status.wb1), and the project budget worksheet (budget.wb1). Three other files are detailed below.

proj_dct.dbf — This is an empty dBASE IV file for use as a project dictionary.

cpia_tsk.doc — This document lists all of the detailed tasks for the Cornucopia project.

pert.bmp — This is an image of a generic PERT chart for the major SDLC activities identified in the Cornucopia project.

monitor.bmp — This is an image of the "Estimated versus Actual" monitoring tables presented in casebook Figure 6-1.

EXTENDED EXERCISES

CIS Lab Extended Exercise 4

Refer to the CIS Lab PERT and Critical Path charts (textbook Figures 6-8 and 6-10) to answer the following questions:

a. What happens to the critical path if the estimate for task 5 is changed to 2.0 time periods?

b. Suppose that in working on task 5 the analyst discovers a serious design flaw, requiring task 2 to be reopened. Would you recommend any changes in the allocation of analyst resources (see the task list in the textbook)? If so, what would you recommend? If not, defend your decision.

c. Redo the PERT and Critical Path charts to reflect the following time period estimates:

Task 2 ... 3.0
Task 5 ... 2.0

Silhouette Extended Exercise 4

Refer to the Silhouette Project Status Worksheet (textbook Figure 6-4) to answer the following questions:

a. Which activities are behind schedule?

b. How accurate were the labor hour estimates?

c. Prepare the "Estimated versus Actual" monitoring tables as of the 8th week.

CHAPTER 7

PROBLEM SOLVING AND DESIGN DEVELOPMENT STRATEGIES

OVERVIEW

This chapter serves as a transition from analysis activities to design activities. The analysis of the enterprise, its existing information system, and its new information needs follows a carefully sequenced pattern of model building. Along the way, the analyst can't help but consider the broad design challenges presented by the new project. Thus, design work really begins during the analysis phase, well before it becomes the focus of the project during the design phase. This suggests that the models used to abstract the existing system, along with the project contract, provide a foundation for the new system design. The modeling matrix presented in textbook Figure 7-5 pulls this all together by showing the correspondence between the five CIS components, the user-driven design activities, and the models. Briefly summarized below, there is a simple recipe for launching the formal design process.

The Transition from Analysis to Design

1. Study the correspondence between the existing system's process, systems, and data models. Identify the major subsystems, user interfaces, and data files. Identify the file processing activities.

2. Study the new system information requirements (problem summary) and objectives itemized in the project contract. Consider each information requirement as a potential subsystem. Consider how each new subsystem might satisfy one or more of the objectives. Consider the broad file manipulation implications of the new subsystems.

3. Compare the existing subsystems with the potential new subsystems, noting the differences and similarities.

The chapter presents several alternative problem solving strategies to consider as you ponder the results of the comparative study discussed above. The list is certainly not exhaustive, nor is it exclusively applicable to systems work. Every analyst develops such a list in either a formal or informal way, which minimizes the need to repeat this particular textbook's list here. However, one overriding theme should be emphasized: Complex problems are best solved in small segments, which are then woven together into well-articulated systems.

Perhaps because there are so many computer solutions available, the analyst may need to conceptualize the new system in terms of one of the three small-enterprise information system stages presented in textbook Figure 7-8 and summarized below.

Small-Enterprise Information System Stages

Stage I	The basic starter system, in which the stand-alone microcomputer is outfitted with a menu-driven solution for master file maintenance, transaction processing, and correspondence. Horizontal software is the centerpiece in such a system.
Stage II	A local area network is employed to facilitate integrated office solutions, including electronic mail, file sharing, and group-based enterprise problem solving and information processing.
Stage III	A wide area network expands the capabilities of information system users to include industry-specific correspondence, and even collaboration, as part of the normal problem solving and information processing activities of the enterprise.

The design prototype is introduced in this chapter as a valuable tool, not only to help the analyst experiment with different solutions, but also as a means of involving the user in the design process. Chapter 12 is devoted entirely to prototyping.

CORNUCOPIA TO DATE

Week 4 of the Cornucopia project finds the analysts virtually finished with the analysis phase and on schedule for the design phase, as illustrated in Figure 7-1 below.

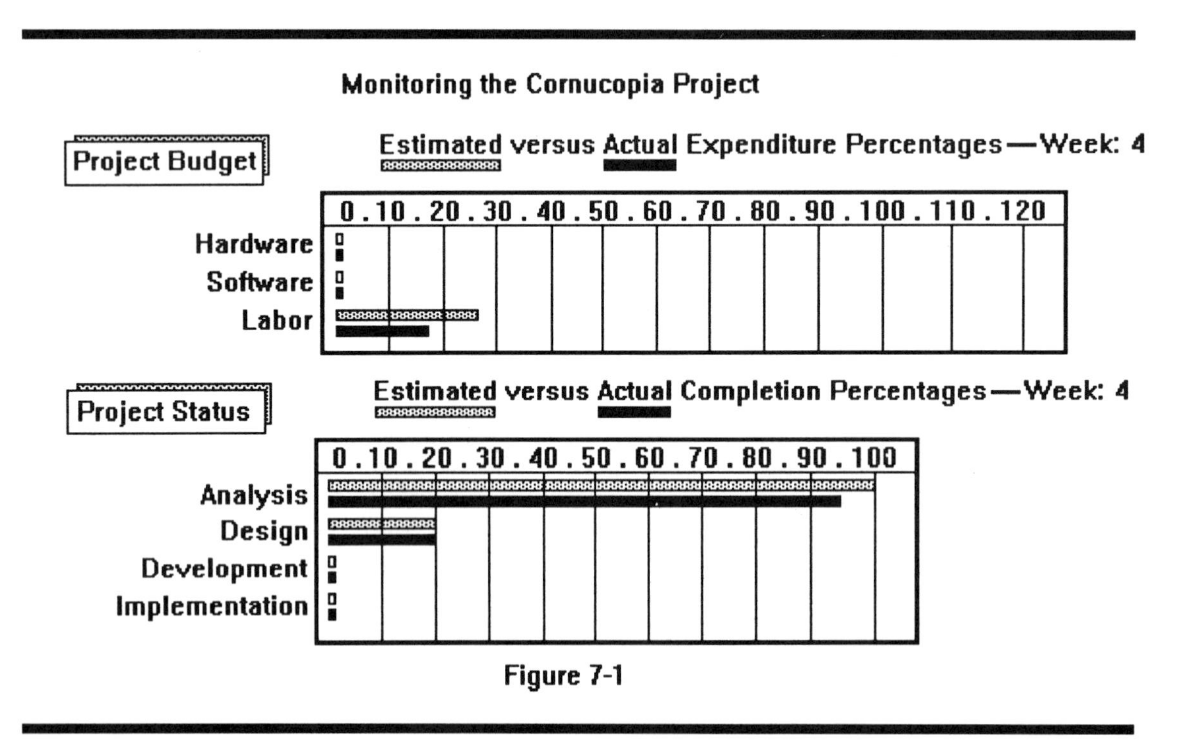

Figure 7-1

The Inside Story

The first-level DFD of the existing system, shown in textbook Figure 4-14, identifies four subsystems. The project contract calls for improvements in four areas, suggesting four subsystems. When these subsystem lists are placed side-by-side (see below), the analysts recognize that the proposed sales trend analysis subsystem has no counterpart in the existing system, and the existing subsystem labeled "inventory" has no counterpart in the proposed system. Several of the subsystems are similar, as are the underlying customer and supplier master files.

Cornucopia's

Existing Subsystems	**Proposed Subsystems**
Customer correspondence	Customer record keeping
Order	Product reordering
Customer correspondence	Customer communications
....	Sales trend analysis
Inventory	....

This exercise confirms the analyst's suspicions that this is a "Stage I" information system project. Thus, they immediately assume that a stand-alone microcomputer and a collection of horizontal software will form the backbone of the information system.

Even though the design details are still to be worked out, the analysts construct the user's system diagram for the proposed system. After deciding on the computer icon to represent a process, the stacked disk icon to represent a data store, and the printer with paper icon to represent the output data flow, the analysts begin with three subsystems: customer file maintenance, CD reordering, and sales trends. This soon leads to the problems summarized below.

Subsystem Processing Concerns

1. Two subsystems (sales trends and CD reordering) require sales transaction data in order to work properly, which adds the sales subsystem.

2. A sales subsystem will require CD master file data, which adds another subsystem, CD file maintenance.

3. The reordering subsystem will require supplier master file data, which leads to a sixth subsystem: supplier file maintenance.

The six subsystems (Cust. Updt., Sales, Trends, CD Updt., Supp. Updt., and Reorder) are now penciled into the proposed system USD, along with the external entity icons from the existing system USD. After some rearranging, the analysts use a paintbrush utility program to draw the USD that appears in textbook Figure 7-14.

How to Copy the Concepts

At this point your biggest challenge is to produce the first draft of the proposed USD, without the benefit of the detailed design work described in the next three chapters. To begin this process, list the major information needs identified in your project contract. For the moment, consider these to be new subsystems. Add to this list any processes that are required to support one or more of the subsystems already listed. Compare this new list to the existing system DFD to determine if some items need to be added or removed. As a point of reference, it is useful to match this rough outline of the new system to one of the small-enterprise information system stages.

Select processing, data store, and output data flow icons for your USD. Beginning with scratch paper, sketch one processing icon for each subsystem and one data store icon for each master file and transaction file in the proposed system. After adding the external entities and at least one output data flow icon for each subsystem, you can begin the process of rearranging your picture to fit your preliminary view of how the new system will fit together. You should expect to make many changes to this view as you work through the design phase activities.

DISK FILES

new_usd.bmp This skeleton image of Cornucopia's new system USD contains six subsystems, three external entities, four data stores, and numerous data flows.

bill_hrs.doc This is a generic billable hours form.

EXTENDED EXERCISES

Sunrise Systems Extended Exercise 1

Review the descriptions and illustrations that introduce Sunrise Systems' Stage II computer information system (see textbook Chapter 7).

a. Prepare a list of the processes, external entities, and data stores that appear in the data flow diagram.

b. Associate the information needs summarized from the project contract with the processes on the above list.

c. Associate the icons used in the user's system diagram with the items on the above list.

d. Explain any discrepancies or omissions you find between the list above and the needs or icons.

CHAPTER 8

FILE AND FORM DESIGN

OVERVIEW

This is the first of three chapters providing detailed instruction on system design. The new system USD is used to start the model-building process for the new design. Through a collaborative working relationship, called joint application design, the analyst and user become closer partners in the problem solving process as they develop each successive model. In small-enterprise work, this relationship is usually informal, but it must be sustained by careful documentation and regular analyst/user communication.

The process, data, and system models can be prepared in the sequence presented below.

New System Model-Building Sequence

1. From the new system USD, sketch the new system DFD.
2. From the new system DFD, sketch the new system ERD.
3. From the new system ERD, sketch the new system GUIDs.
4. From the new system GUIDs, sketch the new system menu tree.

The model-building sequence helps the analyst identify the new system files and file relationships. File design fundamentals begin with an understanding of the common functions served by computer files, two of which are the maintenance of master record information and historical transaction record information. Master files and transaction files, as they are respectively classified, are usually the easiest files to identify in an information system, which explains the textbook's recommendation that the analyst focus on these files first.

Master File Definition

Master files ... describe persons, products, and services that are semi-permanent, or recurring objects of concern to the enterprise.

... are commonly used to record information about customers, clients, employees, suppliers, product descriptions and pricing, service classifications, and so on.

Transaction File Definition

Transaction files	... describe enterprise-related events that occur repeatedly, but usually with different distinguishing characteristics.
	... are commonly used to record information about sales, purchases, meetings, payroll hour exceptions, and so on.

File maintenance screen forms and GUIDs (graphical user interface dialogs) are used to update master files and enter event information into the transaction files. The use of 4GL products to create forms and dialogs greatly enhances the design process. Horizontal database and spreadsheet software include user-friendly utilities that make it easy for the analyst to prototype most of the file processing interfaces. To effectively employ the prototyping utilities, the analyst must first create file structures. Thus, it is difficult to separate file design and form design activities, as revealed in the following activity sequence:

Creating File and Form Prototypes

1. Create a master file structure.
2. Create a screen update form for the master file.
3. Create the GUIDs for the master file update process.
4. Repeat steps 1-3 for each master file and transaction file.

CORNUCOPIA TO DATE

Figure 8-1 summarizes the progress towards project completion.

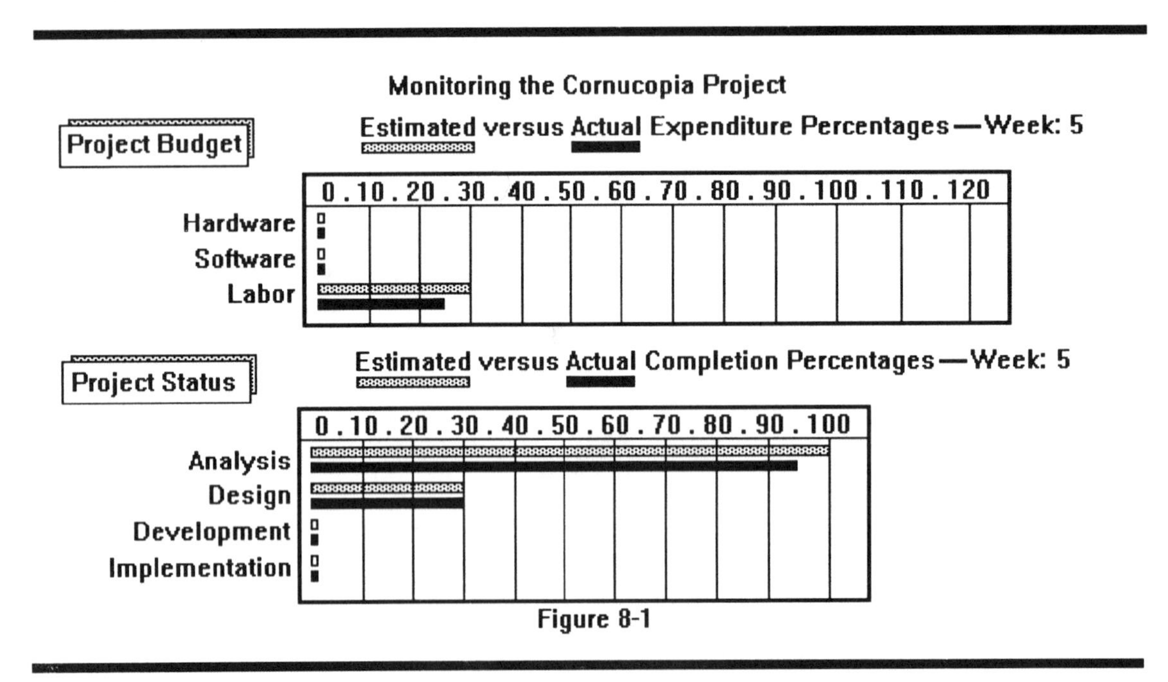

Figure 8-1

The Inside Story

The new system first-level DFD (textbook Figure 8-12) is derived from the new system USD (textbook Figure 7-14), with the following cosmetic changes: The familiar owner, customer, and supplier icons are replaced by the standard external entity symbol, the file icon is replaced by the data store symbol, and the computer icon is replaced by the process symbol. More significantly, the simple connecting lines of the USD are replaced by data flow lines with descriptive labels. Notice that some of the labels are more detailed than others. For example, "old info" generally describes the data that flows from the Customer master file to the Customer Update subsystem, while "date, CD#, sale price" meticulously describes the data flow from the Sales subsystem to the Sales transaction file. This difference is a matter of convenience. The analyst feels that the Customer master file information is standard, and thus, does not require detailed labeling on this diagram. On the other hand, the sales transaction process is of more concern to the analyst, and therefore is more carefully labeled. The "reorder report" data flow is added to correct its omission on the USD.

The analyst creates the new system ERD by placing the DFD's data stores on a piece of scratch paper, along with the file structures. Those files with common fields are connected, with arrows added to define the type of relationship that exists between the files. Because the Customer master file and Form Letter file have a many-to-many relationship (i.e., a single customer may receive several different letters, and a single letter may be distributed to several customers), the normalization process produces another file called Customer/Letters.

The menu tree (textbook Figure 8-15) is created to correspond closely with the subsystems, with several subsystems served by yet another level of menu options. The analyst explains to the user that although this particular menu format may change as the design phase proceeds, it does serve to identify the intended processing options of the new system. Thus, the user can quickly determine if any major pieces are missing.

The analyst initially creates the new system screen forms with the dBASE screen designer utility. As explained in Chapters 12 and 13, the forms that appear in this chapter required a fair amount of code modification (i.e., programming), which will be explained later. More relevant to our discussion here is the similarity among all of the screen images and the way the forms build on one another. This illustrates how the GUIDs work in this project.

How to Copy the Concepts

Use the USD as a guide to create the new system DFD, ERD, and menu tree models. Hopefully, this exercise will give you new insights into the overall design of the new system. It may also frustrate you because your design seems to change with every new model, or you may discover that you don't really understand the new system requirements very well. If this occurs, try to put the frustration aside and concentrate on the master file structures and form design activities for the moment. These tasks will propel the project forward, giving your team some concrete information system products to evaluate and revise. Remember, the textbook's *enhanced* SDLC methodology is presented in a linear format, but in practice you move back and forth between the phases, as you gain a better understanding of the enterprise and more experience with the process.

Identify the master file most easily defined in terms of its structure and use in the new system. Create the file, add five to ten sample records, and use a screen design utility to layout the basic data entry fields needed to update the file. On paper, simulate a dialog between the computer and the user to add, change or delete the master file. This simulation is much like the story-boarding technique used in the film industry. For example, frame number one should present the user with a choice to add, change, or delete the file, along with an option to exit the update. Frame number two should reflect the user's choice from frame number 1. Thus, if the user wants to change a record, frame number two must ask the user to identify the record to be changed by entering the key field value or by browsing through the existing records. Then, in frame number three, the computer should either display the selected record, or inform the user that the record selected by key field value does not exist. You can see that such dialog sequences can become very elaborate, depending on the different possible outcomes to each frame. What should emerge is a pattern to the dialogs, suggesting the basic screen layouts needed to maintain this, and perhaps all, master files in the system. Figure 8-2 presents a partially completed GUID worksheet for this example. Notice that if the user enters an error, frame 3 is displayed, which may lock the user into this sequence forever. The worksheet makes it easier to identify this design flaw, and spot alternative designs (e.g., return to frame 1 or 2).

Once you have gained some experience and confidence in using this technique, you should try to apply it to the transaction files in the system. You are cautioned, however, that transaction files processing can be more complicated than master file processing.

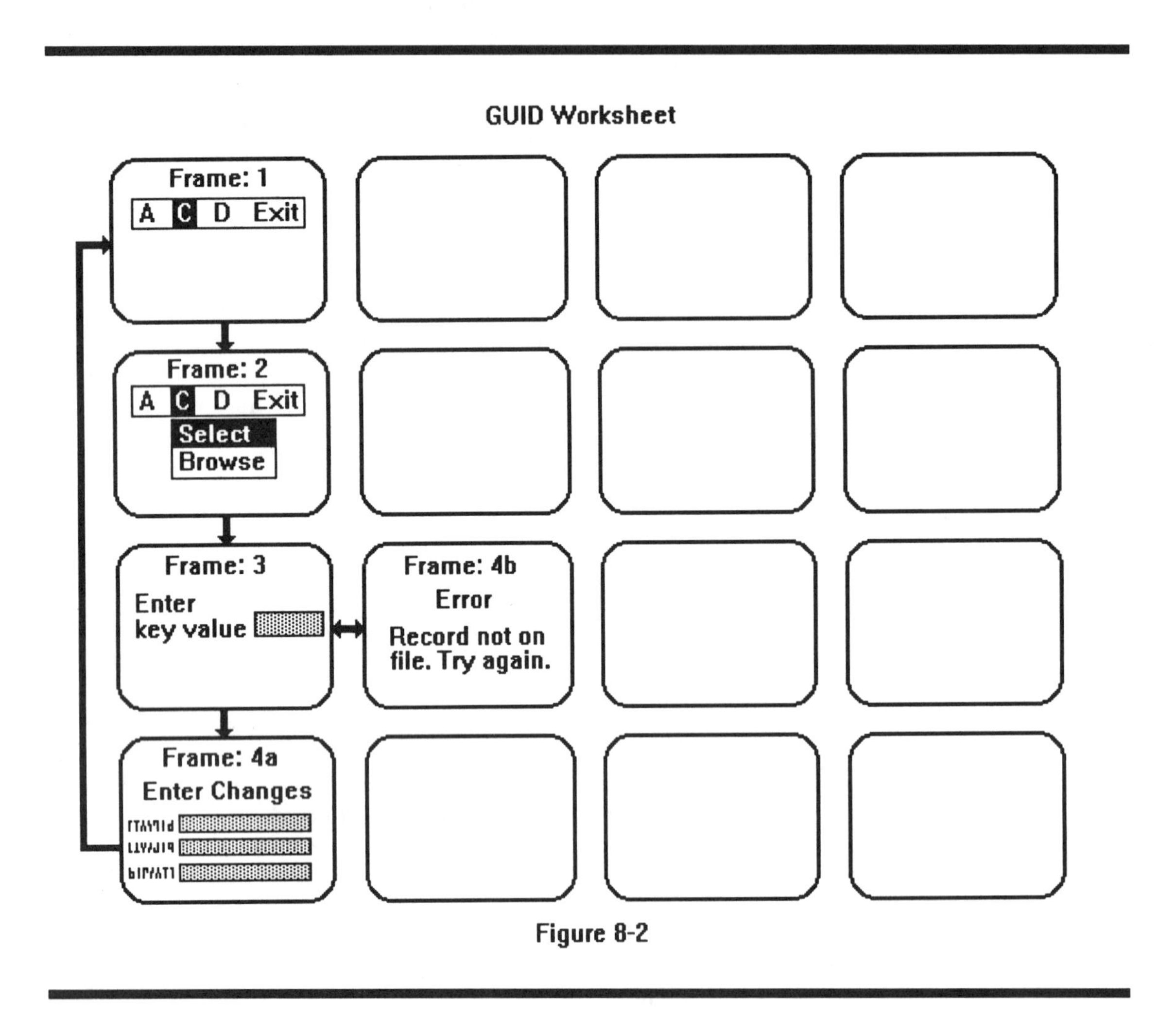

Figure 8-2

DISK FILES

The dBASE IV program files used to create Cornucopia's screen forms are included as part of the discussion in Chapter 13.

guid_wk.bmp This is a blank GUID worksheet.

EXTENDED EXERCISES

Sunrise Systems Extended Exercise 2

Refer to the Sunrise's ERD (textbook Figure 8-7) and database interface template (textbook Figure 8-9) to answer these questions.

a. How is the normalized ERD affected if it is possible for a single investor to have more than one investment portfolio? Develop a new normalized ERD to address the problem.

b. Develop a GUID sequence to handle the Investor master file maintenance subsystem.

c. How does the addition of the new Investor master file to the ERD affect the DFD presented in the last chapter (see textbook Figure 7-11)?

CHAPTER 9

REPORT AND QUERY DESIGN

OVERVIEW

Information system outputs serve many audiences, are generated from many different sources, and follow many different formats. Figure 9-1 presents a worksheet that summarizes these parameters.

Output Worksheet

Title	______	______	______
Audience			
executive	______	______	______
manager	______	______	______
everyday worker	______	______	______
Content			
one source	______	______	______
multi-source	______	______	______
Form			
hardcopy	______	______	______
softcopy	______	______	______
regularly scheduled	______	______	______
on-demand	______	______	______
report	______	______	______
query	______	______	______
user inquiry	______	______	______

Figure 9-1

In practice, users often activate database update screens and forms to view the contents of a file. This is especially true when the database product provides convenient object-oriented file navigation tools, such as scroll bars and buttons. While Chapter 13 discusses object-oriented software more fully, it is important to anticipate how users are likely to adapt the input and output interfaces to suit their needs. The problem with such free lance adaptations is the potential misuse of the interface. For example, since the update form is designed to change the database, users who are only authorized to view file contents, are inadvertently permitted to introduce new records, delete records, and change record values. To avoid problems of this nature, the analyst should design equally convenient output interfaces, along with the proper authorization and file protection procedures.

Offering the user database access through structured query language (SQL) and query-by-example (QBE) is another way to minimize non-standard adaptations. These tools allow the user to design output based on needs that were not explicitly identified during the analysis and design activities. Both SQL and QBE require the analyst to prepare careful user instructions.

CORNUCOPIA TO DATE

Figure 9-2 summarizes the progress towards project completion. The project status percentages show that the analysts have completed the analysis phase and are reasonably close to the design phase estimated completion. The project budget percentages show actual expenditures below estimates for all three major cost elements.

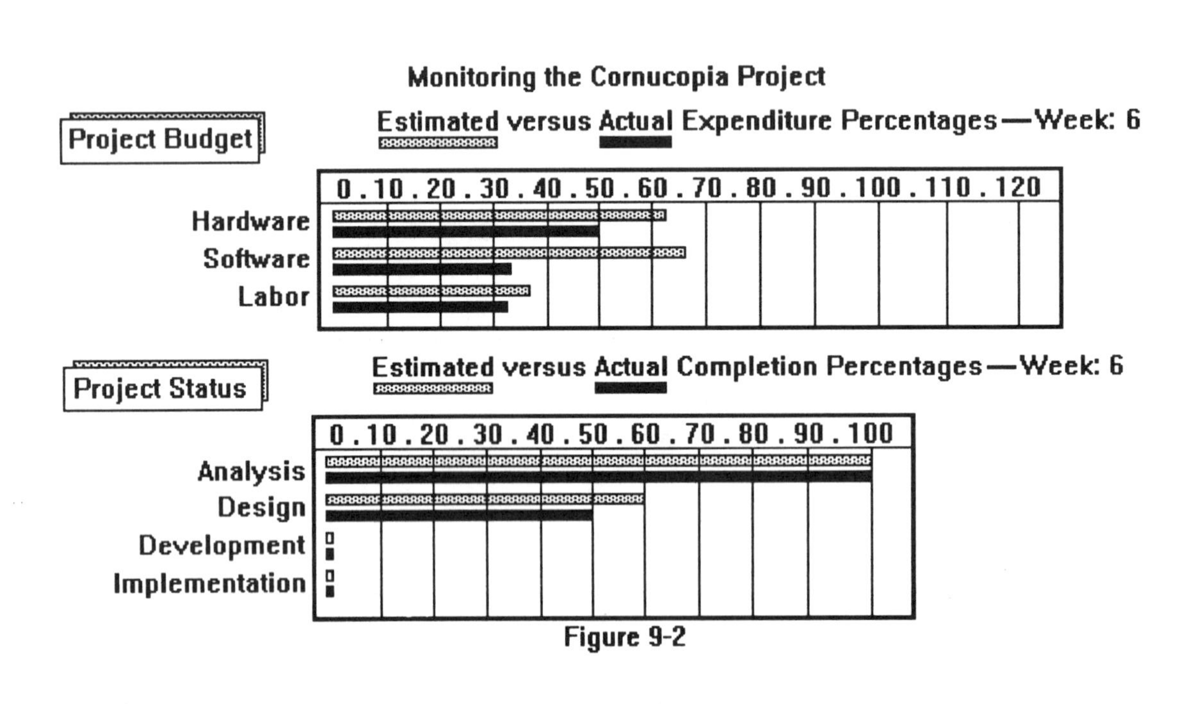

Figure 9-2

The Inside Story

Up to this point, the analyst and user have left the sales trend reports unspecified, even though they generally agreed that these reports would present comparative sales summaries over time. During the recent joint application design sessions, they developed six report and one query layout, as identified in textbook Figure 9-9. Together with the other outputs for master and transaction file maintenance, reorders, and correspondence, there are 16 different items. The menu tree (textbook Figure 9-10) is revised to reflect the specific sales trend reports.

The sales trend output designs show the user's interest in summary unit and amount information. Some outputs separate the information by price category. Some express the information in percentages rather than absolute amounts. After these designs are developed, the analysts inspect the data flow diagram and the entity-relationship diagram to determine if the process and data models can support the output demands. Version 1.0 of these diagrams (textbook Figures 8-12 and 8-13), along with the file structures (textbook Figure 8-14) show a very simple sales transaction processing subsystem. The analysts decide to revise these models to accommodate the sales trend output requirements. The new Daily and Monthly history files will summarize sales transactions in a stair-step fashion -- the Sales subsystem will summarize each day, the Trends subsystem will summarize each month. The revised data flow diagram (textbook Figure 9-11) shows these additions to the process model.

These changes demonstrate how output design influences the entire design process. In short, your system must be able to produce the desired outputs, which is why analysts often begin the design process with detailed output definition. The Cornucopia analysts approach output definition differently because they want to solve the design problems in small stages, beginning with the easiest elements. This allows them to define the master file maintenance tasks early, with the expectation that early output prototypes will facilitate more productive user participation in the design process. Their approach also conveniently illustrates how input, output and process design are interrelated, with new insights in one area leading to revisions in the others.

Hardware and software purchasing begins at this time, even though the information system design is not complete and the analysts have not prepared formal product bid specifications. This is unorthodox, but it is not unrealistic for a project of this size. By now the analysts can safely predict what kind of minimal hardware platform they will need. With a slight upward adjustment to allow for the remaining design decisions, the analyst are assured that there will be sufficient processing resource to support the information system. Further, the relatively short time constraints placed on this project require premature hardware ordering to ensure delivery in time to install and test the new system. Chapter 11 presents a detailed discussion of product procurement.

How to Copy the Concepts

Before any serious output prototypes are produced you should work with the user to develop a standard output template to define standard headings, page sizes, fonts, logos, etc. for both softcopy and hardcopy outputs. Although this activity can proceed somewhat independently of the specific output designs, you must maintain close contact with other analysts to avoid time consuming format revisions.

As with input design, begin your output design activities by concentrating on the easiest products first. Although your original project packet probably contains some sample outputs, these are not necessarily the easiest outputs to produce. Master file reports are good design candidates because they are derived form a single source, they seldom require complex processing, and they can usually be prototyped with report-building software. Once again, early success in generating computer-based products will generate enthusiasm and build team confidence to tackle more difficult output design problems.

Transaction file and complex statistical or interpretive report design should follow master file report design. With these outputs, you should sketch a sample page and then identify where each data element comes from. Some elements may be derived from multiple sources or complex processing algorithms, while others may come directly from transaction file records. In addition to clarifying the output document format and content issues, you will find these annotations helpful during the processing design activities that follow.

The output worksheet is designed to help you document the audience, content, and form issues. You can expand upon this worksheet by adding information regarding special processing requirements, data file ordering or indexing, special user-supplied inquiry inputs, and so on. At a minimum, you should develop a comprehensive list of the output requirements similar to the one illustrated in textbook Figure 9-9.

Finally, after investigating product delivery schedules and the degree of confidence you have with the basic resource requirements, you should consider initiating hardware and software purchases. Given the artificial nature of the team project, you may be inclined to delay this activity on the assumption that you can expect "instant" delivery as late as week 11 or 12 of the project. This approach is not recommended because it rarely works this way in practice.

DISK FILES

out_wksh.doc This is a blank output worksheet.

EXTENDED EXERCISES

Sunrise Systems Extended Exercise 3

a. After reviewing textbook Figure 9-1, develop a departmental level submenu with three output report options. Sketch the design for one of these reports.

b. Considering that in textbook Figure 9-2 the performance section information comes from a spreadsheet , speculate on where the spreadsheet data comes from and how it might be processed.

c. After reviewing textbook Figure 9-7, develop two additional user query outputs -- one periodic and one on-demand.

CHAPTER 10

PROCESS DESIGN

OVERVIEW

Process design is influenced by the number and variety of software options available to the modern analyst. The following summarizes these options.

Processing Software Options

1. Horizontal 4GL products, such as word processors, spreadsheets, and database management programs, require analyst adaptation to satisfy the information needs of the enterprise.

2. Vertical products, such as medical-billing and point-of-sale turnkey systems, require no modifications, but may leave some enterprise information needs unmet.

3. Integrated software, such as Microsoft Works and ClarisWorks, while combining popular horizontal products into one program, still require the analyst to develop enterprise-specific solutions.

4. Software suites, such as Microsoft Office and Lotus SmartSuite, package and price several highly compatible horizontal products as one, but do not eliminate the customizing work of the analyst.

5. Database management code-building software, such dBASE IV's application generator, NOMAD, and ORACLE, significantly reduces the analyst's database-related work, but does not address the other information subsystems.

6. Third-generation language (COBOL) code-building software, such as that included with some CASE tool products, considerably reduces the programmer's workload. Generally, this is not a recommended option for small-enterprise systems work.

7. Completely customized, third-generation language programming, using languages such as C, Pascal, or COBOL, is much too time-consuming to be a practical option for small-enterprise systems work.

This chapter introduces two extremely important computer software developments which transcend all of the options listed above. First, automated file-sharing techniques, such as dynamic data exchange (DDE) and object linking and embedding (OLE), make document-centered processing easier to implement. Second, object-oriented horizontal software products, discussed in more detail in Chapter 13, promise to transform file processing into object processing, further reducing the need for detailed programming. A summary of file-sharing methods appears below.

File-Sharing Methods

Low-tech	Paper-based cut and paste with scissors and tape Computer clipboard-based cut and paste using a mouse or keyboard Keyboard or macro-based file import and export
High-tech	Dynamic data exchange Object linking and embedding

Finally, this chapter presents the annotated subsystem flowchart and the subsystem structure chart, both of which help the analyst specify process designs. The annotated subsystem flowchart shows the processing option and corresponding file types the analyst intends to use to implement the process design. The subsystem structure chart shows how the functional components of a software product work together. The progression of process design charting techniques presented in the textbook is summarized below.

Process Design Charting

1. Menu tree
2. Composite system flowchart
3. Annotated subsystem flowchart
4. Subsystem structure chart
5. Program flowchart

CORNUCOPIA TO DATE

Figure 10-1 summaries the progress towards completion.

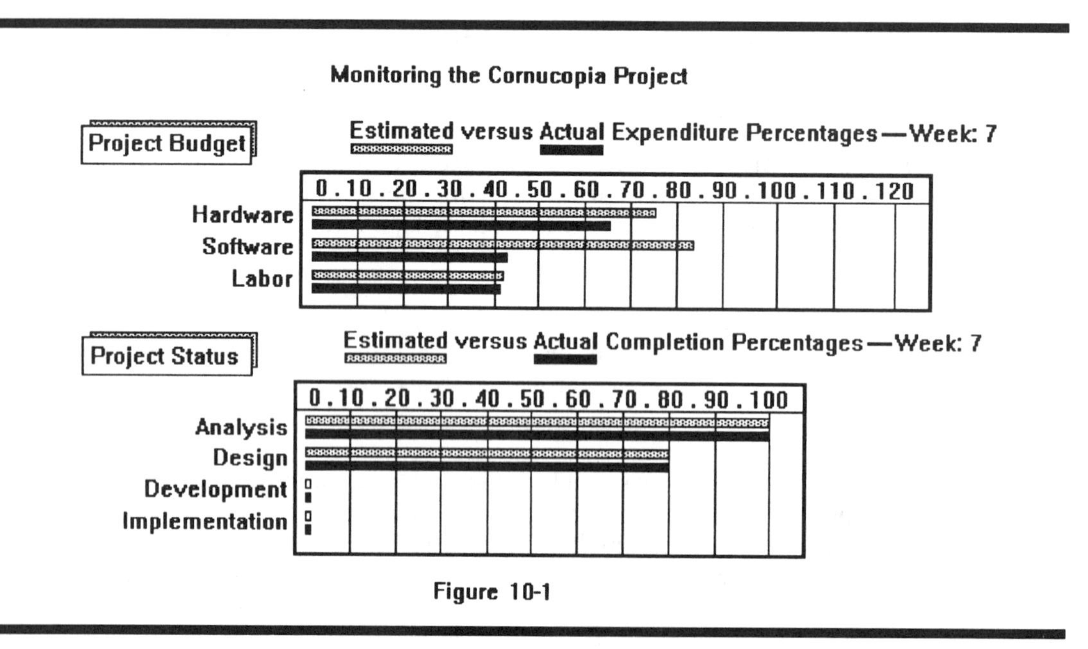

Figure 10-1

The Inside Story

The broad processing design parameters are determined by the information needs of the enterprise, which are well documented at this point in the project. The analysts first detailed decision concerns which processing option or options they intend to employ. As stated previously, the vertical solution available through Phono-Scan is rejected for obvious pedagogical reasons. The programming-intensive solutions are also rejected because they require significant course prerequisites. This leaves the horizontal software option, which is consistent with the basic premise of the textbook: Using *enhanced* SDLC methodology, student-analyst teams can construct small-enterprise computer information systems from microcomputer-based hardware and software resources.

Choosing specific horizontal software for the Cornucopia project was driven by very practical concerns for software availability in the typical educational setting and academic preparation of the typical student. While object-oriented products are likely to dominate future systems work, a traditional relational database product (dBASE IV) was chosen as the primary horizontal product because it satisfies both practical concerns. The discussion in Chapter 13 includes object-oriented products to demonstrate what may soon become standard.

The annotated subsystem flowchart presented in textbook Figure 10-17 reveals the importance of the database processing in this project. Database files or processing is involved in every subsystem. As with input and output design, the analysts begin detailed process design with the master file maintenance subsystems. To accommodate the screen form designs and the interface dialog sequences, the analysts must design a dBASE program. The subsystem structure chart presented in Figure 10-18 reflects the analysts' first attempt to identify the various functional modules of the program. Unlike its revision in Chapter 13, it does not conform to structure chart construction standards.

How to Copy the Concepts

After you develop the basic subsystem definitions, you must decide on which horizontal products to use. Your practical concerns parallel those described above. Namely, you are limited to the products available in the computer lab environment and your level of product-specific expertise. If you choose dBASE as the database product, you will be able to adapt the master file maintenance programs included on the student resource disk (see Chapter 13). The transaction file processing code is provided as well, but to satisfy your particular project requirements, you will probably need to develop your own code for this subsystem.

If you have access to an object-oriented database management product, such as Paradox for Windows, you should seriously consider working with this powerful technology. Admittedly, learning a new product while you are learning an implementing methodology like the *enhanced* SDLC is a real challenge. However, the rewards for such effort are substantial.

Once you have selected the horizontal software for your project, you should construct the annotated subsystem flowchart. This chart is extremely important. It is used to identify the process prototypes, programming needs, and file-sharing requirements. It can also be used to assign analysts to work on different portions of the development work that lies ahead.

Develop the structure charts in as much detail as possible, but don't worry if they seem very general at this time. Your experience with upcoming process prototyping and development activities will help you to refine these charts.

DISK FILES

struct.bmp This skeleton image of Cornucopia's sales transaction subsystem structure chart contains several functional modules.

EXTENDED EXERCISES

Sunrise Systems Extended Exercise 4

a. Compare Sunrise's composite system flowchart and annotated subsystem flowchart (textbook Figures 10-4 and 10-5). Are all the files, screens, and reports represented on both charts? If not, make the necessary corrections.

b. Modify the annotated subsystem flowchart to accommodate an investment analyst's on-demand query output regarding portfolio performance.

c. Use the annotated subsystem flowchart to identify the source files and source applications for each of the three portions of the OLE output document illustrated in textbook Figure 10-8.

CHAPTER 11

COST/BENEFIT ANALYSIS

OVERVIEW

Working on the assumption that every project accumulates costs and provides benefits, the analyst prepares a cost/benefit chart to provide some economic justification for the project. The analysis involves identifying both tangible and intangible costs and benefits.

Cost elements are developed in a manner consistent with the delineation of the five components of a computer information system. The following summary compliments the worksheets that appear in the textbook.

Project Cost Elements

Tangible Cost Components

- Hardware
  - (platform, monitor, printer, etc.)
- Software
  - (operating system, system utilities, backup, security, etc.)
- Data
  - (file preparation, transmission costs, etc.)
- People
  - (training, maintenance, etc.)
- Procedures
  - (documentation, etc.)
- Analyst
  - (labor, travel, supplies, etc.)

Intangible Cost Components

- Risk
  - (complete failure, product liability, etc.)
- Opportunity
  - (alternative investment, etc.)
- Morale
  - (employee alienation, customer dissatisfaction, etc.)
- Obsolescence
  - (functional, operational)

Project benefits are developed by assigning dollar values to the measurable objectives itemized in the project contract. Cost avoidance may be considered a tangible benefit as well. The following list summarizes the discussion in the textbook.

Project Benefit Elements

Tangible Benefits
- Contract objectives
- Cost avoidance

Intangible Benefits
- Improved organizational flexibility
- Improved enterprise image
- Improved employee morale
- Improved product and service quality

Once the costs and benefits are itemized, the analyst plots the cumulative amounts on a chart. When cumulative costs are surpassed by cumulative benefits, the new system has "paid for itself", so to speak. There are other methods for evaluating the economics of computer information system projects, but the simple cost/benefit method is usually sufficient for the small-enterprise environment.

This chapter completes the "Design" section of the project. Accordingly, the "Design Review Session" is scheduled to formally present the new system input, output, and process design, resource requirements, and cost/benefit analysis.

CORNUCOPIA TO DATE

Figure 11-1 presents the progress towards completion, which points out that the project is slightly behind schedule with respect to development activities.

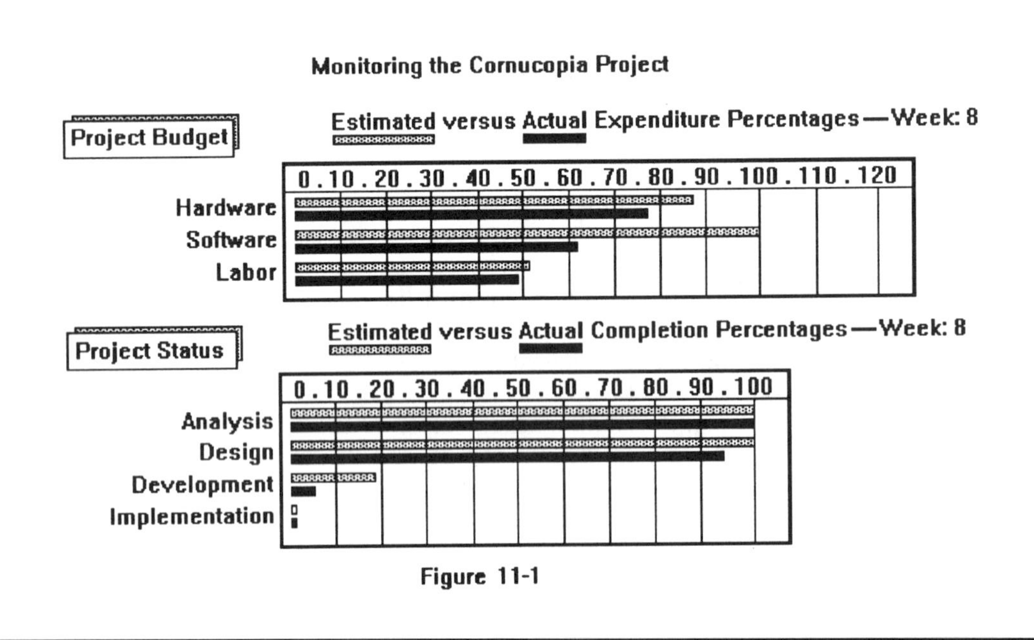

Figure 11-1

The Inside Story

Cost/benefit analysis often occurs earlier in the project. In this treatment, it is delayed to coincide with the scheduled completion of the design activities. Along with the detailed resource specifications, the analyst has a better understanding of how the new system will meet the objectives identified in the project contract. Thus, the estimating tasks may be somewhat easier to complete.

Given textbook publication lead-times, Cornucopia's resource specifications are obviously out-of-date. This situation is unavoidable in this framework, but easily remedied in practice. Your resource specifications and cost figures should reflect the changes that have occurred in the industry and the most recent pricing.

The analysts evaluated each project objective to estimate the tangible benefits. These estimates are based on specific assumptions about the value of the owner's time and sales patterns in the store. Although these numbers are speculative, the user agrees with the assumptions and is willing to monitor her time savings and sales volumes to provide the data necessary to evaluate the estimates.

The analysts' actual labor hours expended (42) compares favorably with the estimated labor hours (44), but the distribution of these hours is of some concern. Compared to the estimates, analysts spent less time on analysis and more time on design. The analysts and user wonder if the labor hour estimating method is correct.

How to Copy the Concepts

To develop a complete project cost estimate you must first prepare a detailed resource specification. Most of the resource requirements were itemized during the input, output, and design activities. Assemble a list of these items and compare it to the sample in textbook Figure 11-9. Make adjustments to your list as needed. The next step is to associate prices with each item. The Request for Bids, which is a standard approach to this task, is probably unnecessary for your project, especially if you have access to popular computer magazines and catalogs.

Although the analyst labor estimates are already established, you may be tempted to adjust some of the numbers to reflect your experience with the project. Such adjustments are not recommended in this setting, but if they are approved, be sure to indicate the revisions on the project budget and status report.

Perhaps the most difficult part of the cost/benefit analysis is translating the project objectives into specific benefit amounts. This requires some guess work, along with whatever experience you may have with the enterprise's operations. Be sure to document the rationale used to make the guess. In this way, you can improve your estimating ability by adjusting your method as you gain real-world experience.

As mentioned in Chapter 1, a cost/benefit spreadsheet template is included on your resource disk (cost_ben.wb1). In the broadest sense, you can follow the Cornucopia example to build this chart by simply substituting your project benefits for Cornucopia's.

DISK FILES

cpia_res.doc This is the detailed list of resource requirements for the Cornucopia project (textbook Figure 11-9).

EXTENDED EXERCISES

Sunrise Systems Extended Exercise 5

a. Update the Request for Bids (textbook Figure 11-3) to reflect the latest high-end microcomputer products.

b. Prepare a detailed hardware and software resource requirements list for Sunrise.

c. Prepare a list of potential tangible and intangible benefits associated with Sunrise's networked solution.

CHAPTER 12

PROTOTYPING

OVERVIEW

In earlier chapters, prototyping is presented as a design activity. In this chapter, prototyping is presented as a product development activity. In practice, prototyping overlaps both activities, making it difficult to distinguish between design and development.

There are several types of prototypes, as summarized below.

Types of Prototypes

Reusable	The prototype is used as the basis for the final product. Prototyped input/output interfaces and processing logic are expanded to include menu accessing features, error detection, file accessing logic, inter- and intra-module communication mechanisms, and so on.
Throwaway	The prototype is used to facilitate design and user participation, after which it is discarded. The analyst, relying on the basic prototyped design parameters, develops the real information system with different tools.

With each of these prototypes, the analyst can build different levels of information products, as summarized below.

Levels of Prototypes

Input/Output	The prototype is confined to the input and output interfaces, with no access to the underlying processes and files.
Processing	The prototype expands the input and output interfaces to include access to the underlying processes and files. Usually, this prototype is restricted to simple processes, such as master file maintenance.
System	The prototype provides menu access to all of the subsystem functions, including transaction file processing. Often times, the size of the data files and range of transaction values is restricted.

When the analyst uses modern 4GL design products, prototype construction is a natural consequence of the design process. Database management products, equipped with screen and report builder utilities, provide one of the most useful prototyping tools. Spreadsheet software is used to create highly computational information products, while word processing packages are used to construct textual products. System software utilities and graphics software are used to fashion menu interfaces. Each of these tools permits the analyst to develop computer-based design products that can be used to develop the final, operational product.

The analyst must be aware that prototypes, so easily created with these tools, can sometimes mislead both the analyst and the user about the difficulties of developing the full-featured products. Also, there is some danger that the analyst and user may mistake a pretty prototype for hard-nosed product documentation.

Throughout the chapter the analyst is referred to as an analyst/programmer. This is to alert you to the likelihood that the development phase will include some programming activities, regardless of the extent to which the system has been prototyped.

CORNUCOPIA TO DATE

Figure 12-1 presents the progress charts. Notice that the development phase remains behind schedule.

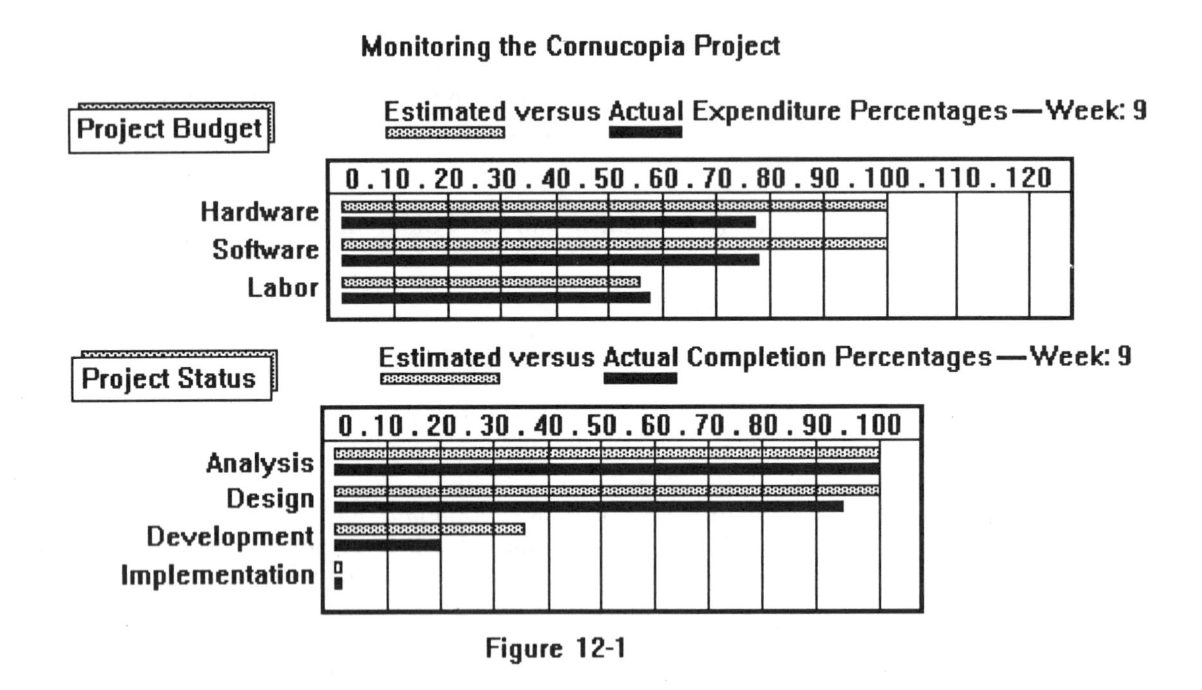

Figure 12-1

The Inside Story

With the exception of some very early menu formats (see textbook Figure 8-15), the analysts create reusable, input/output and processing prototypes for this project. The first basic screen form and graphical user interface dialog designs are sketched using the Customer master file maintenance subsystem processing requirements as a guide. As discussed in Chapter 8, the analyst must define master file structures (i.e., field names, data types, key fields) before prototyping the file maintenance interfaces. To reiterate, this sequence is required because the analysts are creating relational database prototypes, rather than stand-alone graphic images.

Thus, prototyping begins well before system design is complete, which explains why the file structures, screen forms, output displays, and even menu options change several times. Also, 4GL prototyping tools, such as database screen and report builders, encourage the analyst to actually begin the development process during the design phase. This is exactly what happens with the Cornucopia project, as illustrated by the dBASE code that supports the master file maintenance interfaces (see textbook Figures 12-14 through 12-16).

Once the database file is created, the analyst uses dBASE's forms designer utility to create a basic screen design and generate the associated dBASE code. Textbook Figure 12-15 illustrates how the analyst uses the command editor to simplify the generated code by eliminating machine supplied documentation and dBASE environment settings. The analyst will add these two items back during the development phase, when these program modules become part of the file maintenance subsystem documented in textbook Figure 10-18. Notice that the programming effort is reduced because all three programs use the same screen locations, data element names, user prompts, and picture clauses. Another module (see Chapter 13) produces the upper portion of the screen.

As prototyping overlaps into product development, system design is naturally solidified. At the same time, the analyst team turns its attention towards tasks that require a mixture of 4GL application development and conventional computer programming. This tends to narrow the analyst's focus, leaving fewer opportunities for the free-form problem solving activities that characterize the design phase.

How to Copy the Concepts

Because this discussion formalizes prototyping activities that span the entire design phase and portions of the development phase, you have already implemented some of the methods presented in the textbook. If you choose to use dBASE as the implementing relational database product, you can begin to create your prototypes by simply modifying the master file maintenance code supplied on the student resource disk. Cornucopia's transaction file processing and output report prototypes can serve as a model for your project, but you may need to develop completely new programs or at least make significant changes to the Cornucopia code.

On the other hand, if you use an object-oriented database product, your prototyping experience will be slightly different than the one described for the Cornucopia project. With object-oriented products, screen forms and reports can be closely associated with file processing through individual data element attribute and method specifications. Thus, it is more likely that your input/output prototypes will incorporate significant portions of the file processing design. Also, your interface screens may include a wide variety of navigational and operational icons, such as buttons and scroll bars. This allows you to produce user-friendly access and processing features without any programming effort on your part.

DISK FILES

All of the Cornucopia database files and programs are included on the student resource disk. A complete listing appears in Chapter 13.

EXTENDED EXERCISES

PRC Extended Exercise 1

Review the context diagram, data flow diagram, system flowchart, and output prototypes for the Political Research Corporation (PRC). See textbook Figures 12-5 through 12-8.

a. Correlate the images on the data flow diagram and system flowchart to identify the following match-ups:

data flow diagram:	system flowchart:
processes	4GL software
data stores	file types
data flows	input/output documents

b. Identify the master files and transaction files on the system flowchart.

c. Describe the tools you would use to create the prototype products identified on the system flowchart and illustrated in textbook Figure 12-8.

CHAPTER 13

4GL PROGRAMMING

OVERVIEW

The combination of automatic code generation and goal-directed programming commands is commonly called 4GL programming. This chapter addresses the programming options open to analysts as they work to transform prototyped products into working information systems. Collectively, these options reflect an evolutionary trend that promises to make programming much more accessible to professional programmers, analysts, and even users. The following summarizes those options that are particularly relevant to small-enterprise systems projects.

Programming Options

Third-generation language	These languages (e.g., COBOL, Pascal, C) require the talents of a highly-skilled programmer to fashion procedural commands into a series of complex programs.
Fourth-generation language	Using analyst supplied design specifications, these languages (e.g., ORACLE, NOMAD, Genifer) automatically generate code that resembles a third-generation language program.
Object-oriented language	These languages (e.g., C++, Visual Basic) require the talents of a skilled programmer to integrate application-specific objects with object and method libraries to create highly customized information systems.
4GL product	These products (e.g., dBASE IV) provide easy-to-use menu options, as well as nonprocedural and procedural commands, which the analyst can assemble into programs. Macro record and automatic code generator features simplify the task.
Object-oriented product	These products (e.g., Paradox) permit the analyst to create user-friendly processing and file access interfaces by combining application-specific objects with a rich library of objects and methods.

Regardless of the programming option, the analyst-programmer must plan the sequence of programming commands, object manipulations, and dialogs required to produce prototypes and ultimately, the real information system. As summarized below, there are several tools the analyst can use to develop such plans.

Program Planning Tools

Modular structure chart	Identifies the sequential, repetitive, and conditional relationships among program modules.
Program flowchart	Identifies the sequential, repetitive, and conditional relationships among individual program commands.
State-transition diagram	Identifies the dialog-dependent actions of the information system.

CORNUCOPIA TO DATE

Figure 13-1 presents the status and budget progress charts. In order to get back on schedule with respect to project development, the analysts devoted 20 labor hours during the week. This brings the cumulative labor expenditures to 81% of the total budgeted.

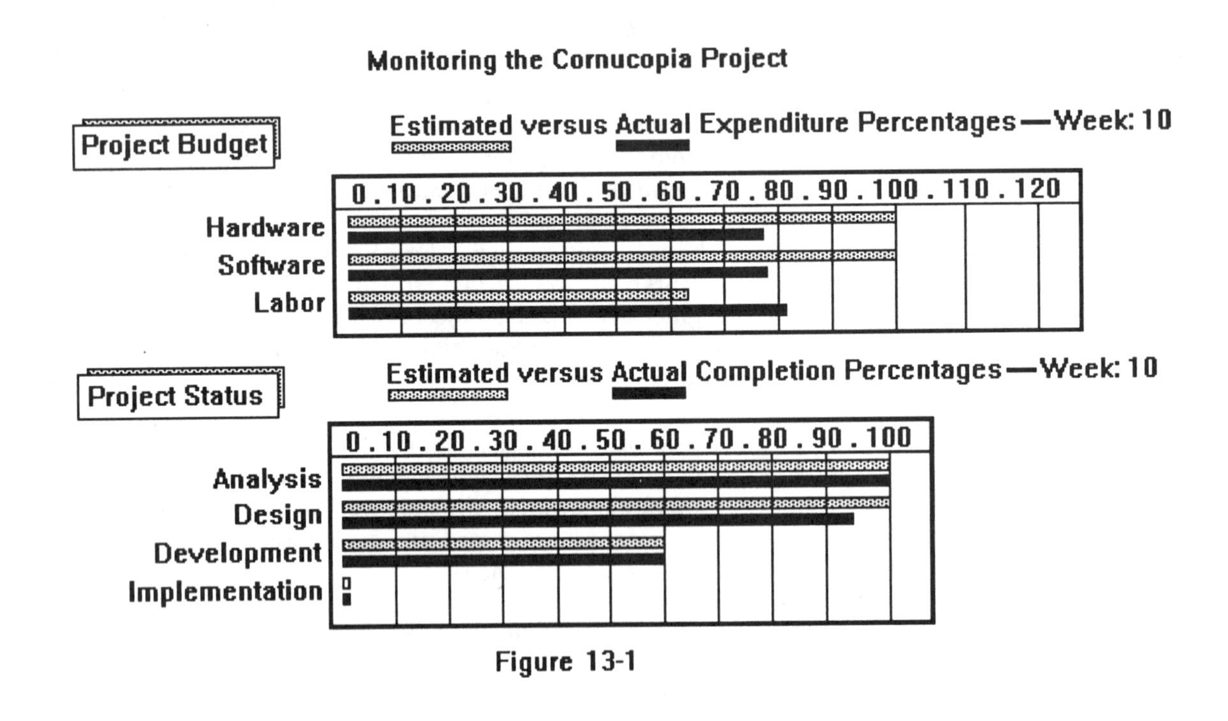

Figure 13-1

The Inside Story

The decision to use dBASE IV is based on the fact that this product enjoys a long history of successful use and is readily available to students and educational institutions. Although this DOS-based product includes screen and report designer features and an automatic code-generating utility, the analysts create several program modules to enhance the master file maintenance interfaces and develop the sales transaction subsystem. Much of this programming is routine and serves as a model for student projects based on similar DOS products.

Even with all of the code included in this chapter, several elements of Cornucopia's new information system are missing. The information system reports and queries are not implemented, leaving only the prototypes illustrated in textbook Chapter 9 for your reference. The correspondence subsystem is only partially developed, with a single mail-merge operation and macro illustrated in textbook Figures 13-21 through 13-23.

The spreadsheet application presented in textbook Figures 13-24 through 13-26 adds a completely new element to the system. This is included to illustrate a practical spreadsheet adaptation and alert you to the possibility that the analyst can sometimes accommodate small system design enhancements during the development phase of the project.

The Windows version of dBASE promises to be different because of its object-oriented nature. Other Windows-based, object-oriented database products (e.g., Borland's Paradox and Microsoft's Access) are available as well. Textbook Figure 13-27 is included to illustrate how the Cornucopia project might employ an object-oriented product to access CD sales history.

How to Copy the Concepts

As demonstrated by the inflated labor hours charged to the Cornucopia project during the programming activities, you should expect to devote a considerable amount of time to product development. With your focus now on programming activities similar to those described in this chapter, it is imperative that you finalize the basic system design in order to minimize the need to make time-consuming program changes later. One way to accomplish this is to call a special team meeting to discuss the design, agree on development priorities, and assign task responsibilities.

Task assignments should be based on analyst skills. It is likely that one or two team members will possess the skills required to complete the programming required. Once the overall design is agreed upon, the remaining team members can work concurrently on other development phase tasks. Word processing, spreadsheet, and graphics specialists can concentrate on developing information products that the programmers can later incorporate into the final product. Also, project documentation and management activities can be assigned to nonprogramming analysts.

At times it may be beneficial to assign a nonprogrammer to act as a sounding board to programmers who are bogged down. Such a team often discovers that a fresh mind invigorates the problem-solving process. Extending this practice throughout the task assignments helps to guard against a situation in which team members are isolated from one another. Regular, open communications among analysts is very important, not only to avoid inconsistencies in product development, but also to reduce the risk associated with a prolonged analyst absence and territorial jealousies. Do not permit team members to completely withdraw from the group, no matter how convincingly they may pledge to return with a finished product.

DISK FILES

Your resource disk contains all of the files necessary to run the Cornucopia database applications illustrated in the text. All but one of these files are collected together in subdirectories, as noted below. If you display these directories, you will notice that the disk also contains index (.mdx) and object (.dbo) files associated with the database (.dbf) and program (.prg) files. To execute the database programs: (1) Place the student resource disk in drive A, (2) Launch dBASE IV, (3) Escape to the dot prompt, (4) Set the default drive to A (type SET DEFAULT TO A: <Enter>), (5) Set the default directory to A:\ (type SET DIRECTORY TO A:\ <Enter>), and (6) Execute the Cornucopia application (type DO MAINMENU <Enter>).

\mainmenu.prg

\custmain

customer.dbf	**blankcus.prg**
custmain.prg	**custform.prg**
custadd.prg	**browcust.prg**
custchg.prg	**savecust.prg**
custdel.prg	**restcust.prg**

\cdmain

cd.dbf	**blankcd.prg**
cdmain.prg	**cdform.prg**
cdadd.prg	**browcd.prg**
cdchg.prg	**savecd.prg**
cddel.prg	**restcd.prg**

\saletran

sales.dbf	**sales.prg**
dayhist.dbf	**upsales.prg**
saletran.prg	**salesum.prg**
blanksal.prg	

EXTENDED EXERCISES

PRC Extended Exercise 2

Review the flowcharts, diagrams, and object-oriented illustrations in textbook Figures 13-2, 13-3, 13-4, 13-7, 13-8, and 13-9.

a. Notice the copyright at the bottom of the ADDVOTE.PRG (textbook Figure 13-2). What does this suggest about ownership of custom software?

b. Notice the double vertical lines used in the program flowchart processing symbols labeled DO BLANKVOTE and DO VOTEFORM. What do these double lines mean?

c. How would you change the entity-relationship diagram (textbook Figure 13-3) to reflect the possibility that a voter can participate in more than one survey?

d. Design an object-oriented form to access an individual survey response.

CHAPTER 14

SYSTEM ENVIRONMENT

OVERVIEW

Computer hardware, system software, 4GL software, and customized software combine to create the system environment. This chapter summarizes the major hardware and software issues facing the small-enterprise analyst. Considering the pace of technological change in the computer industry, it is virtually impossible to keep up with every innovation. One alternative to this dilemma is to focus on those factors that influence hardware and software performance and analyst productivity the most. The following summarizes the major hardware concerns for either the "PC" or Macintosh solution.

Hardware Issues

Processor	The microcprocessor sets the broad performance constraints of the system. In addition to the fundamental throughput characteristics (speed, register width, etc.), the range of future upgrade options is of critical importance.
Memory	To execute efficiently, modern software requires large amounts of memory. Generally speaking, extra memory improves software performance by reducing the amount of disk access required to swap program segments and manipulate file records.
Disk	Disk size and access speed are important because of the large increase in system and application software size.
Bus	The width of both internal and external communication pathways influences the performance of the hardware components itemized above.

The analyst should match operating system software capabilities with information system requirements. For example, the decision to adopt an operating system with a graphical user interface, or networking capabilities, or high-speed data transfer should depend upon the degree to which users are expected to interact with the operating system, the anticipated network requirements, and the data transfer frequencies and volumes.

The analyst must install the information system products on the computer hardware-software platform. In other words, the user needs a convenient procedure to launch the information system. One approach, often referred to as a turnkey system, is to set the system start-up procedure to default to the information system. Although this is an easy way for the user to get into a particular information system, it makes it more difficult for the user to directly access the hardware-software platform or any other installed application software. Another approach is to install the information system in a way that allows the user to launch any number of applications, utilities, or information systems via the standard operating system interface.

The chapter includes a broad introduction to network technology. The following checklist recaps the discussion in the textbook.

Networking Checklist

To establish the physical connection you need:
- a) a connecting medium.
- b) network interface cards.
- c) a modem for WAN access.

To establish the logical connection you need:
- a) a network operating system.
- b) a dedicated server or peer-to-peer protocol.
- c) network-friendly application software.

To integrate the technology into the work place you should consider:
- a) how to best facilitate worker collaboration.
- b) whether or not mobile computing is appropriate.

CORNUCOPIA TO DATE

Figure 14-1 presents the status and budget charts.

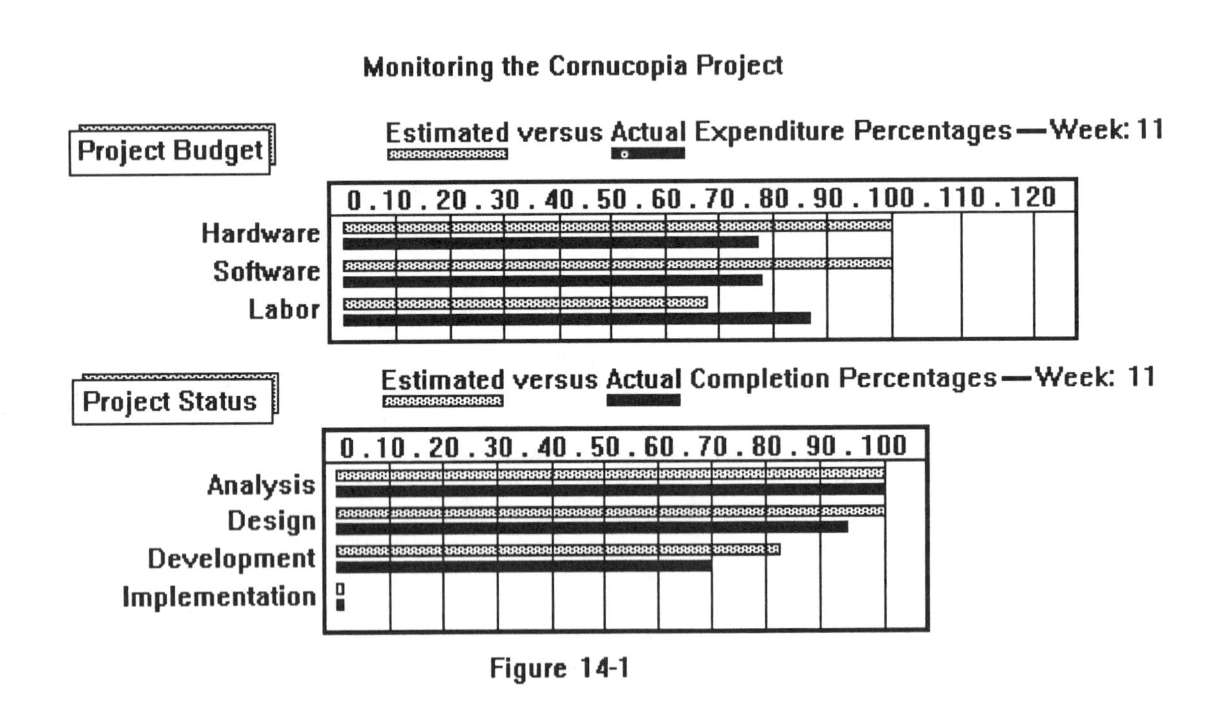

Figure 14-1

The Inside Story

The textbook discussion of the Cornucopia project repeatedly mentions possible design enhancements, such as networking, data communications, and point-of-sale technology. To implement these elements, analysts must have access to sophisticated hardware and software, posses the necessary technical skills, and have the appropriate amount of time to design and develop the products. Because such access, technical skill, and time may not be available to most student teams, the Cornucopia model project is restricted to a simple, single-station information system. Likewise, the choice of a PC and Windows hardware/software platform for Cornucopia is predicated on the assumption that these very popular products are most likely available in student computer labs.

The decision to employ the standard Windows interface to implement the Cornucopia menu options means that user access to the final product differs slightly from the menu tree (textbook Figure 9-10). Also, it may be difficult to associate the subsystems appearing in the user's system diagram (textbook Figure 7-14) with the program items in the Cornucopia program group. Certainly the analysts could use a batch file program or a menu-building utility program to create a more accurate implementation of the design. However, the simple Windows solution is employed because it is more easily duplicated by students.

How to Copy the Concepts

Your project team's decisions about hardware, software, and product installation are driven by practical considerations. In most settings, you are restricted by your computer lab's hardware and software resources, the skill level of team members, and the length of the academic term. As stated before, the project approach to learning relies heavily upon hands-on experience and an appreciation for the process we refer to as the *enhanced* SDLC. This means that it is better to actually go through the experience rather than simply talk about the experience, even if it means compromising the design to accommodate the restrictions imposed by the simulation. It also means that it is extremely important that you recognize these compromises, so as to distinguish the simulation from the real thing.

To put the above admonition another way, your team must attempt to deliver at least some of the information system products required for your project. Then, your team must carefully document those elements of the delivered system that do not work or do not match the design specifications. In this way, you will learn that although we strive for perfection in systems work, we must remember that absolute perfection is an illusion. Every system can be improved. Our goal here is for you to experience that process.

DISK FILES

There are no files on the student resource disk that refer specifically to this chapter.

EXTENDED EXERCISES

PRC Extended Exercise 3

a. After reviewing textbook Figure 14-7, develop some plausible entries for the Program Item Properties required to include all of the subsystems identified in PRC's data flow diagram and system flowchart (textbook Figures 12-6 and 12-7).

b. Develop the hardware specifications required to install PRC's information system on a PowerPC Mac.

c. Research the most recent prices for the PRC network upgrade described in textbook Figures 14-13 and 14-14.

d. Develop the upgrade specifications for a dedicated server network.

CHAPTER 15

SYSTEM TESTING

OVERVIEW

The chapter title is meant to describe a project-long series of testing procedures. Beginning with program- or application-specific syntax testing and concluding with user acceptance testing, these procedures become increasingly more complex. Incremental testing allows the analyst-programmer to minimize testing complexity by limiting the number of features and functions that are being tested at any one time. Thus, when a test fails, the analyst-programmer can generally assume that the error is caused by the new feature or function being tested, with a commensurate reduction in the scope of the debugging effort.

The following summarizes the various types of testing required during the SDLC.

Types of Testing

Design phase	
User interface	Tests the interface design for basic operational error checks, such as data type validity, reasonableness, and consistency.
Development phase	
Program module	Tests the subsystem program modules for syntax, logic, and run-time errors.
Integrated module	Tests the subsystems to determine if the program modules can work together as designed.
System	Tests the system to determine if the subsystems and main menu can work together with system hardware and software.
Pilot	Tests the system to determine if it can function with real data values and volumes, system procedures, and system personnel.
Implementation phase	
User acceptance	Tests all five system components to secure formal user acceptance of the system.

During user interface testing and much of development phase testing, the analyst-programmer creates fictional test data. The goal is to test as much of the system as possible, without compromising the budget or the project completion time-table. Accepting the fact that it is impossible to test every single combination of data values or processing situation, the analyst-programmer documents exactly which system features and functions are tested by fictional test data. In this way, the analyst-programmer builds a reusable set of data that can be adjusted to add, delete, or amplify testing sequences.

Beginning with the later stages of system testing, live data is used. To some extent, this randomizes testing in terms of data values, sequences, and volumes, which is valuable given the limitations mentioned above. Once again, it is important to document the circumstances under which data is captured and testing occurs. This reduces the effort required to plan future tests dictated by system maintenance and upgrades.

The best way to document system testing procedures is to develop a testing plan, which includes well-defined schedules, user and analyst-programmer responsibilities, and target goals. This plan is especially useful to future analysts, who may not be familiar with the information system or the necessary testing procedures.

CORNUCOPIA TO DATE

Figure 15-1 presents the status and budget charts, which reveal a reassuring 90% completion for the development phase activities.

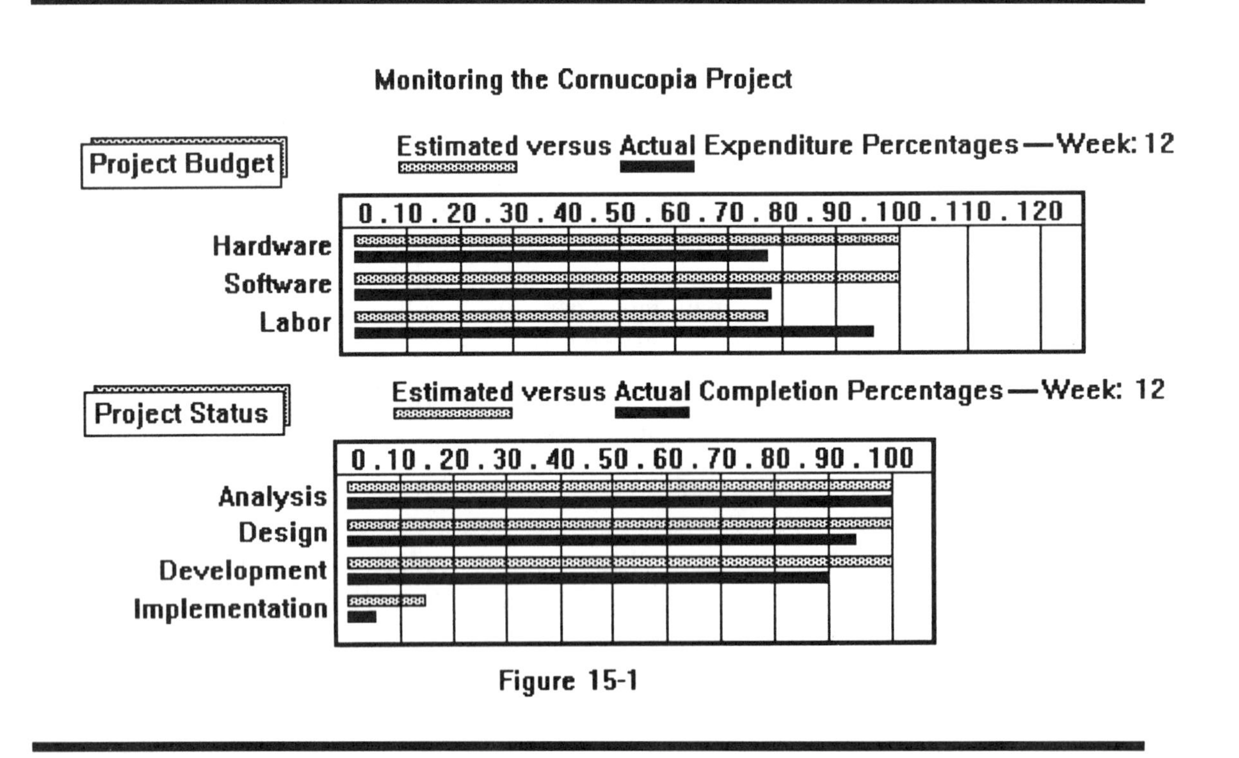

Figure 15-1

The Inside Story

User interface and program module development and testing for the database applications produces routine syntax, logic, and run-time errors. As mentioned previously, the Customer master file maintenance subsystem is developed and tested first, which produces the basis for the CD master file maintenance subsystem. Because there are so many different possible interface scenarios, a small testing matrix, illustrated below, serves to document master file user interface testing. Each line in the matrix requires a different sequence of user inputs. The comment entries summarize the expected outcome.

Master File Interface Testing Matrix

menu selection	search method	action	continue?	comment
ADD	***	U(pdate)	N	update, clear screen & return to menu
ADD	***	U(pdate)	Y	update, clear screen & continue
ADD	***	C(hange scr)	*	allow changes to screen values
ADD	***	A(bort)	*	clear screen, flag record & ask "continue?"
CHANGE	Browse	***	*	if EOF, return to menu
CHANGE	Select	***	*	if no match, display error & continue
CHANGE	Select	***	*	if match, display record & allow changes
CHANGE	***	U(pdate)	*	update, clear screen & return to menu
CHANGE	***	C(hange scr)	*	allow changes to screen values
CHANGE	***	A(bort)	*	clear screen, restore record & rtn to menu
DELETE	Browse	***	*	if EOF, return to menu
DELETE	Select	***	*	if no match, display error & continue
DELETE	Select	***	*	if match, display record & allow changes
DELETE	***	D(elete)	*	update, clear screen & return to menu
DELETE	***	A(bort)	*	clear screen, restore record & rtn to menu

While a similar matrix can be established to guide the sales transaction interface testing, the mail-merge and sales analysis functions do not lend themselves to this approach. The analysts are disappointed to find that these functions do not perform their file sharing tasks very well, and they are difficult to use. The analysts agree that it would be better to use more compatible horizontal software, such as a software suite or an integrated package.

As described in the textbook, integrated module and system testing of the master file maintenance subsystems presented a small challenge with respect to screen management. This results in some awkward code revisions. The analysts agree that an object-oriented database product might offer a more "elegant" solution. In reality, this product never undergoes pilot or user acceptance testing. Once again, we are reminded that this is a sample project, and thus, not subject to the rigors of the real world.

How to Copy the Concepts

The initial interface and program module development and test cycle can be assumed by the individual analyst-programmer assigned this task. This can actually reduce development and testing time by eliminating the need to explain simple problems that are best remedied by referring to software manuals or code samples. In other words, at this stage, the analyst-programmer often needs uninterrupted work time to become proficient with a piece of software. This approach is reversed during the later testing stages, when subsystems are fit together and installed into a new system environment. At that time, the small team approach is useful if only because it brings the primary analyst-programmer out of what may be a somewhat isolated perspective.

Your team should develop a project-wide plan to implement a systematic testing sequence, assign testing responsibilities, and set goals for error tolerances and completion dates. All during the testing activities you should take note of those products and procedures that are well-suited to the prototype review session. You may find that some interfaces, processing modules, and operating procedures are too complex or confusing to include in your presentation, while others are extremely straight forward. In addition, be aware that product testing may generate a great deal of information that you can use to document your information system. For example, the testing matrix presented above could be adapted to serve as an operations matrix.

DISK FILES

testctl.doc This is the product testing control sheet presented in textbook Figure 15-7.

testproc.doc This is the product testing procedure form presented in textbook Figure 15-7.

EXTENDED EXERCISES

PRC Extended Exercise 4

a. After reviewing PRC's testing plan (textbook Figure 15-7), develop a "Product Testing Procedure", including test data and expected results, for the billing screen form.

b. Describe, in general terms, the system testing procedures for the new PRC main menu options.

c. Describe how you would involve the user during pilot testing of the client billing upgrade.

CHAPTER 16

SYSTEM DOCUMENTATION AND TRAINING

OVERVIEW

System documentation provides a detailed history of the project, which current and future users and analysts refer to during the long period of system maintenance and review. The numerous and varied documents, illustrations, file specifications, diagrams, etc. are collected together into a project binder during the analysis, design, development, and implementation phases. In preparation for system training, conversion, and installation, the analysts creates three manuals from the materials in the binder. The content and purpose of each of these manuals is summarized below.

Product Documentation Manuals

Training Manual	This manual is used to guide user training. It contains training objectives, schedules, system overviews, exercise outlines, quick reference guides, and copies of training session visual aides.
Procedures Manual	This manual instructs enterprise personnel on how to operate the information system. It does not describe how the information system products are integrated into the overall operation of the enterprise. It contains a description of the information system, operating instructions, input/output samples, emergency and security instructions, error messages, plus numerous system resource specifications and definitions.
Reference Manual	This manual provides detailed technical information about file layouts, special formulas, processing table values, and special codes.

The analyst must be sensitive to the audience, the training detail requirements, and the training setting. Some audiences require overview information, others require very detailed instruction. Some training takes place in the laboratory, some is conducted in the work place. However, one perspective remains constant: Good training relies on a carefully monitored training cycle that includes training, feedback, adjustment, and retraining.

The chapter presents two training methodologies. Instructor-directed learning prescribes a predefined training cycle scenario. In this case, the teacher plans each exercise to include specific activities and achieve specific objectives. An alternative, user-directed, approach retains learning objectives similar to the instructor-directed approach, but leaves much of the activity planning open to the user's discretion. Given the high degree of user participation in the project, it is natural to adapt and integrate portions of both approaches to fashion a set of CIS training techniques. The techniques summarized below are designed to establish a good foundation using formal implementation-phase training sessions and then encourage informal user exploration and discovery.

CIS Training Techniques

Instructor-directed	Lectures, demonstrations, hand-on exercises, and tutorials are used to guide the user through a progression of learning experiences that develop a general understanding of the information system, as well as introductory-level operational skills.
User-directed	Problem-solving situations and unstructured exploration sessions provide an informal learning environment that exploits the user's curiosity and encourages user initiative.

The chapter concludes with an important discussion on computer ethics. Considering the steady stream of ethical dilemmas presented throughout the text, the ACM and DPMA standards offer some objective information to guide analysts' and users' behavior.

CORNUCOPIA TO DATE

Figure 16-1 presents the latest updates to the project status and budget percentages. Notice that although the project is currently on schedule, actual cumulative labor charges already exceed the total estimated labor charges for the project.

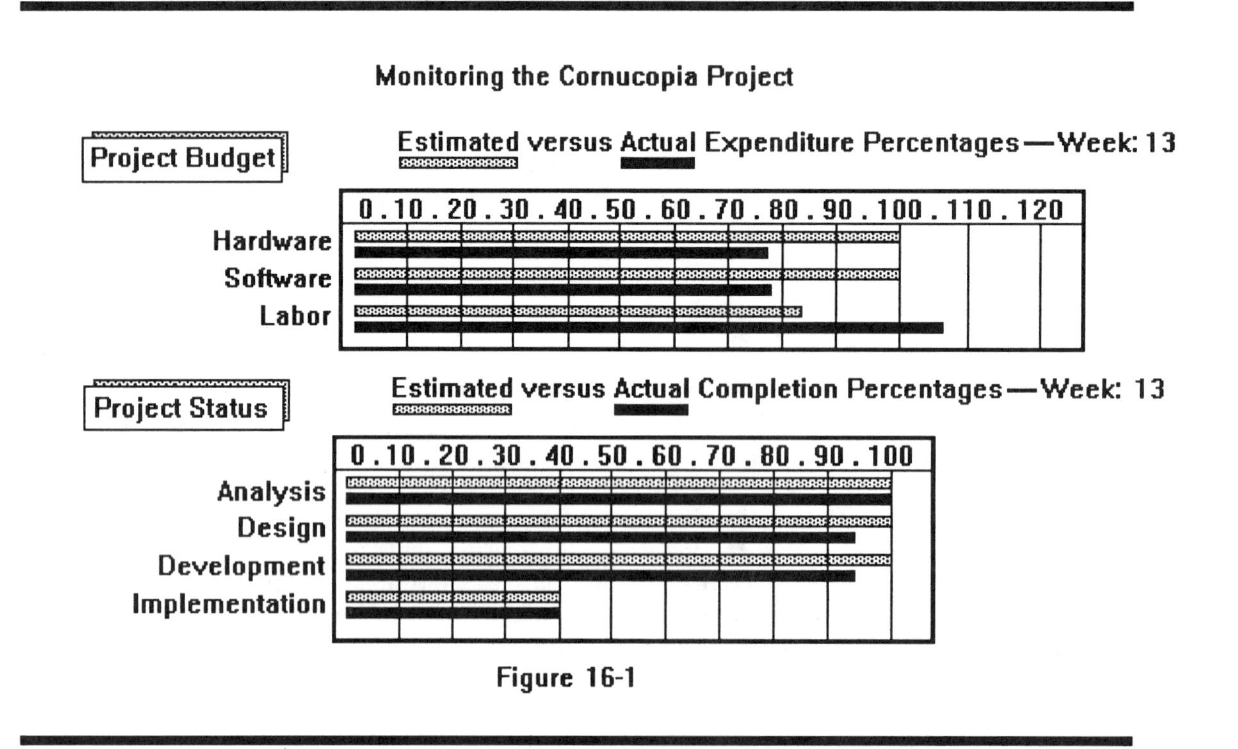

Figure 16-1

The Inside Story

Through casual observations and informal conversations, the analysts discover that the owner wants to retain primary responsibility for the new computer information system. This requires different levels of training for the sales clerks and the owner. As indicated in textbook Figure 16-11, three training sessions are designed for all employees and one session is reserved just for the owner.

Due to the nature of the enterprise and attention devoted to customers, the analysts and owner feel that the initial training sessions should occur after normal business hours. To minimize the inconvenience to employees, the analysts agree to conduct training on-site, which means they would normally need to transport the hardware from their office to the business for each session. Fortunately, there is an empty office at the rear of the store, so the computer can remain on-site after the first training session. This requires the analysts to complete their computer work on-site.

While most of the initial training is instructor-directed, towards the end of the third session, users are exposed to some problem-solving exercises. Two follow-up sessions provide an opportunity for analysts to introduce more user-directed learning activities.

In all cases, the analysts are careful not to overwhelm the users with technical jargon. Rather, they concentrate on practical training activities designed to build operational skills and general knowledge of the system. The product documentation manuals provide ample technical detail for those users who feel the need to investigate the system further.

How to Copy the Concepts

Every interaction between the user and the information system should be observed to determine the areas of emphasis and content levels of the forthcoming training sessions. Beginning with the first joint application design meetings and continuing through the prototype review session, you should note those elements of the system that are confusing or difficult for users to navigate. These notes, filed away in the training section of the project binder, are the genesis of lecture material, hands-on exercises, tutorials, and general problem-solving sessions.

Assuming that your team will have a limited amount of time for the in-class training session, you must decide whether you want to cover one subsystem extensively or several subsystems briefly. Which ever choice you make, be sure to inform the audience of your strategy. This prepares the audience for the level of detail you intend to present and help to reduce their anxiety about the time constraints.

Rehearse your training session. This includes equipment set-up, software launching, visual aide positioning, lighting to minimize screen glare, presentation timing, and any other aspect of the session that might influence its outcome. In practical terms, this is probably your last opportunity to impress your classmates and teacher, or your real-world clients.

DISK FILES

None of the files on the student resource disk refer specifically to this chapter.

EXTENDED EXERCISES

LandScapes Extended Exercise 1

Refer to the USD, hands-on training exercises and ethical issues presented in textbook Figure 16-8 through 16-10.

a. Research and recommend hardware resource specifications for *LandScapes'* notebook computer.

b. Develop training exercises to teach users how to generate customer billing output and reports.

c. Develop three more ethical scenarios for *LandScapes'* users to consider.

CHAPTER 17

SYSTEM CONVERSION

OVERVIEW

The orderly transition from one information system to another can only be achieved through careful planning, attention to detail, and well coordinated effort of analysts and users. Some of the later stages of system testing and much of the training activity coincides with site preparation, installation, and file conversion.

File preparation involves converting existing files to new formats and creating new files from scratch. New file creation can be very time consuming and often requires the expertise of enterprise personnel. Their experience helps decode source documents, identify errors, and resolve data value conflicts. Many times existing files can be electronically converted to new formats. Although this shortens the file preparation task, there may be new fields to enter or data value verification activities. Again, this work is best left to people who are familiar with the data and who have a real stake in its accuracy. Both types of file preparation are usually performed while the enterprise continues to operate and thus, continues to generate changes to the files. Analysts must develop and implement a plan to capture and process these ongoing changes to the new files. In this way, the old and new files are synchronized.

The chapter presents three different project conversion options as summarized below. Each option generates different costs and risks.

Conversion Options

Direct	This is a low cost, high risk option in which the new system replaces the old system all at once. Costs are low because there is no additional expense to operate two systems at the same time. Risk is high because there are no easy backups in case of new system failure.
Phased	This is the moderate cost, moderate risk option in which the new system gradually replaces the old system. Costs are moderate because of the confusion surrounding the mix of old and new procedures. Risk is moderate because new system failures are restricted to one or two subsystems at a time.
Parallel	This is the high cost, low risk option in which the new system operates in tandem with the old system. Costs are high because of the duplicate effort required to run two systems. Risk is low because you can almost guarantee the new system will not fail by comparing the performance of both systems before converting to the new procedures.

The small-enterprise environment offers no special advantage to any of these conversion options. Each project must be evaluated in terms of existing enterprise workloads, system complexity, user computer experience, and user attitudes about change.

A good conversion plan can reduce both costs and risk. The plan should include a description of the conversion option, a schedule of activities, and a detailed list of analyst and user responsibilities. The Gantt chart is an excellent tool to document these last two items. Textbook Figure 17-5 illustrates the extent to which many of these activities cross over the somewhat artificial boundaries established by the different phases of the SDLC.

Project review concludes the analysts' day-to-day involvement in the project. It provides a definitive end to the analysts' responsibility to fulfill the terms of the project contract. Any further commitments, other than standard product guarantees, should be covered by a new contract focusing on maintenance and review activities. There are two important components to the project review. A systematic evaluation of the product, which should be documented, and a formal document declaring the user's acceptance of the product.

CORNUCOPIA TO DATE

Figure 17-1 presents the latest updates to the project status and budget percentages.

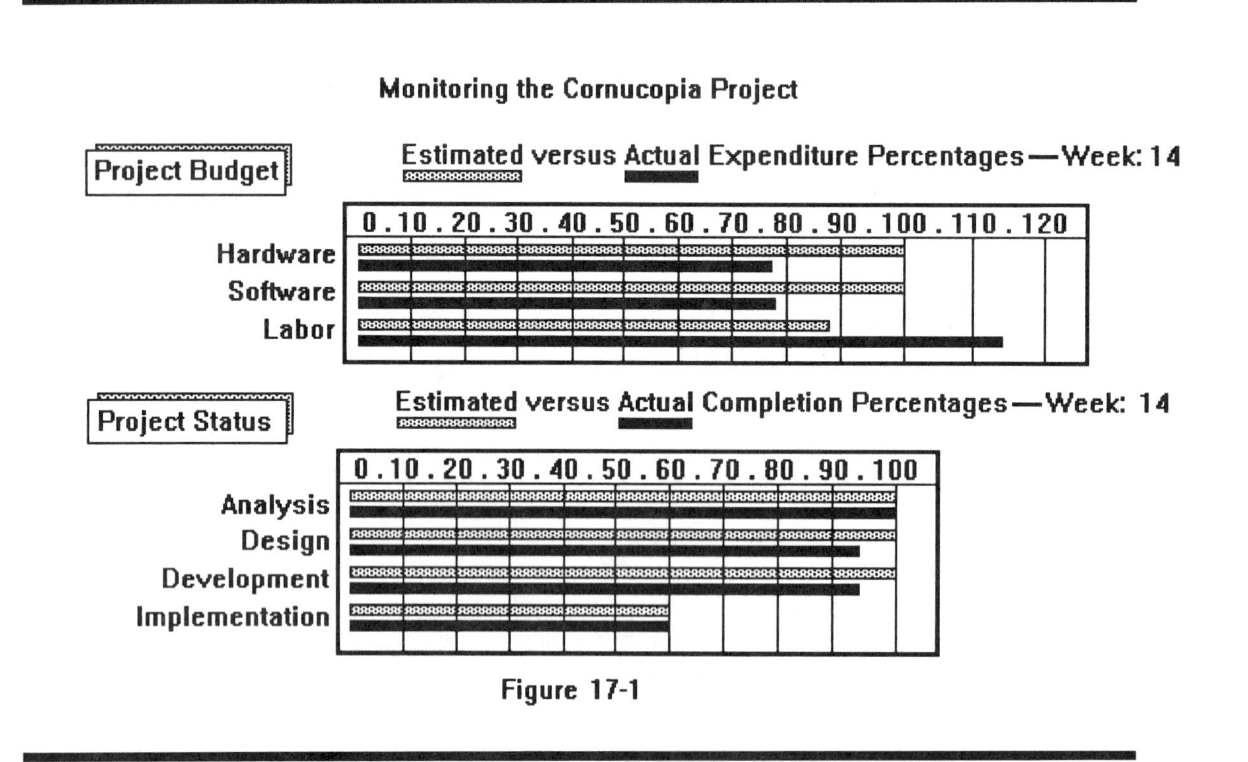

Figure 17-1

The Inside Story

Of the three major files (Customer, CD, Sales) and two ancillary files (Dayhist, Mohist), the analysts only need to prepare the Customer and CD master files during the conversion activities. The initial new Customer master file is created by transforming the existing Apple II file into a dBASE file using the file import feature. Several new fields are entered to complete this process. On the other hand, the CD master file must be built from scratch. This is a big job, considering the number of CDs in the present inventory. The analysts decide to shorten this task considerably by purchasing an electronic version of the CD master file. Further, they make the unusual decision to pay for this file by reducing their fee. This reflects a small accommodation for the excess labor charges recorded during the development phase.

The direct conversion option is attractive because it keeps costs to a minimum and involves almost no risk. The limited scope of the information system protects essential sales transaction processing and cash register operations from any potential new system failures. Even if the new computer refuses to boot up, the enterprise can continue to serve its customers, record sales, and order CDs.

Predictably, the project review goes well, especially since the owner is forgiving with respect to the juggling act performed between the hardware, software, and labor budget accounts. The analysts properly respond to the owner's eagerness to add new features to the system by suggesting a careful analysis of the request, while everyone has an opportunity to adjust to the new system.

How to Copy the Concepts

For the most part, the team project approach breaks down towards the end of the SDLC. It is difficult to simulate file creation, conversion and synchronization because there are no existing files to worry about in this artificial environment. As you have discovered, you must create all of the files from scratch. Likewise, the conversion option decision is never implemented, which makes the exercise somewhat contrived. In spite of these problems, you should proceed with the decision making and documentation process as if you were going to carry them out. There are numerous advantages to this exercise, but the most important point is that you will have a complete project binder to refer to when you do, in fact, need to complete a real project.

Thus, your conversion plan should be sufficiently detailed to guide you and the user through this complicated part of the project, even if some elements of the plan are highly fictionalized. Notice that the analysts chose a Gantt chart, rather than a PERT chart, to document the conversion plan. This choice reflects an awareness of the way the plan is used. Both analysts and users participate in the conversion activities. Therefore, it is necessary to use a tool that both groups will easily understand. Even if you assume that the user has an above average amount of experience with computer systems, the Gantt chart is preferred because it will require less explanation.

The project review is another conversion activity that requires a make-believe treatment. One way to make this seem more real is to ask each member of the project team to review the project. If possible, you might also solicit comments from other project teams.

DISK FILES

review.doc This is a sample project review form.

EXTENDED EXERCISES

LandScapes Extended Exercise 2

a. After reviewing textbook Figure 17-1, create from 5 to 10 sample records for each of the four files. Be sure to coordinate the key field values where appropriate.

b. Review *LandScapes*' detailed Gantt chart (textbook Figure 17-5) and the project milestone events (casebook Figure 1-1). Overlay the scheduled design review session, prototype review session, training session, and final report onto the Gantt chart. Reconcile the timing and content of these snapshot-like presentations with the ongoing, overlapping activities itemized on the Gantt chart.

CHAPTER 18

INFORMATION SYSTEM MAINTENANCE AND REVIEW

OVERVIEW

This chapter identifies three categories and several subcategories of information system maintenance, as summarized below.

Information System Maintenance

Corrective	Analysts are called upon to fix interface, processing, and logic errors discovered after the initial flurry of cut over errors.	
	Cosmetic	These errors affect interface appearance and operational convenience, but are not otherwise of immediate concern.
	Nonfatal	These errors require immediate attention, but do not cause the entire information system to stop working.
	Critical	These errors cause the system to halt.
Routine	Analysts and users are required to adjust and service the information system on a regular basis, even when there are no apparent problems with system operations.	
	Preventive	The user cleans, adjusts, and monitors the system on a regular schedule.
	Adaptive	The analyst is called upon to change the system, without making major changes to any of the system components.
Upgrade	In response to user requests to modify the information system, the analyst initiates a restricted sequence of analysis, design, development, and implementation activities. This is a continuation of the existing system's SDLC, not a new SDLC.	

Programmed review is a regularly scheduled and systematic evaluation of all facets of an operational information system. The subsequent programmed review report often identifies the need for adaptive or upgrade system maintenance. Sometimes, it identifies the onset of system obsolescence. In order to make such assessments, the analyst must develop and apply performance norms against which the existing system can be measured.

The following summarizes the first symptoms of functional obsolescence.

Symptoms of Functional Obsolescence

Costs exceed benefits	This symptom appears when the combination of system maintenance and upgrade costs exceed system benefits.
Needs exceed resources	This symptom appears when the existing system resources can no longer be upgraded to accommodate the changing information needs of the enterprise.

The chapter concludes with a reminder that computer professionals also require regular "maintenance" in order to keep pace with the fast-changing industry. Currency training should be a part of your routine.

CORNUCOPIA TO DATE

Figure 18-1 presents the final updates to the project status and budget percentages. As expected the project completes on-time and slightly under-budget.

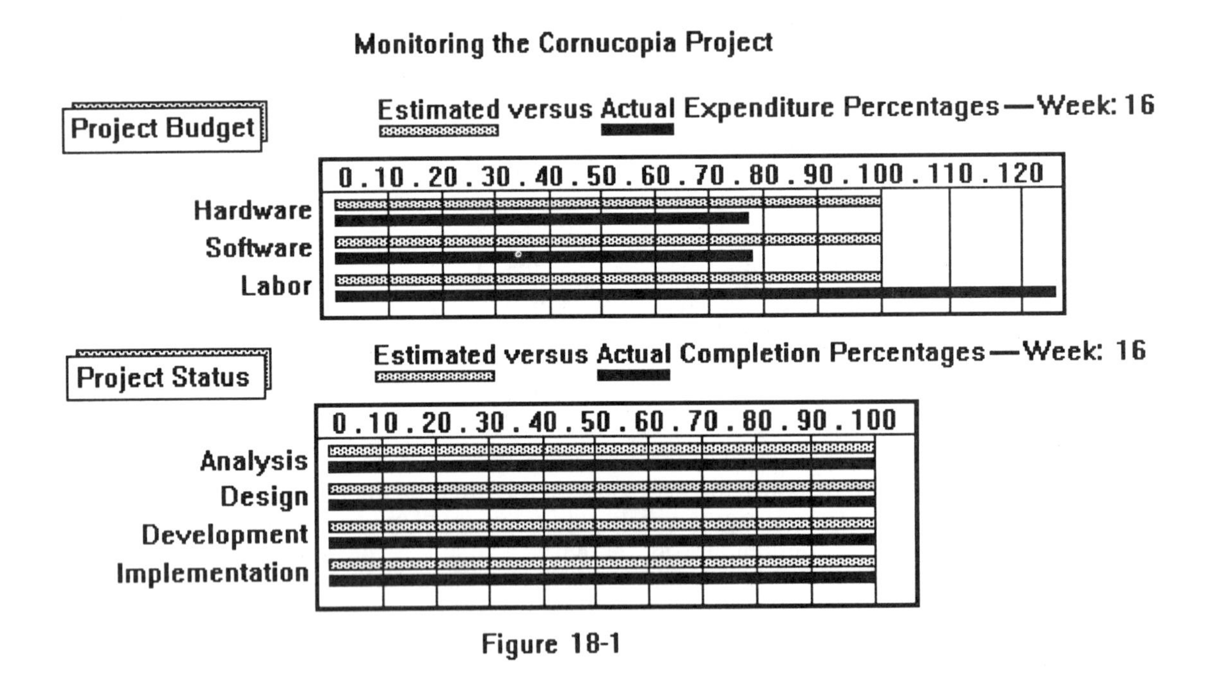

Figure 18-1

The Inside Story

The owner's rejection of plans for formal program reviews and the maintenance contract illustrates the independence of the modern user. In general, this independence is fostered by users' increasingly sophisticated exposure to computing as a part of their everyday life. Also, the participatory nature of the *enhanced* SDLC gives users a sense of self-confidence that encourages them to think through computer problems rather than seek help.

The system upgrade plan offers a glimpse of a more full-fledged information system for Cornucopia. As stated before, this sample project is purposely restricted to make it parallel potential student projects of similar complexity. In fact, the point-of-sale, turn-key system described in textbook Chapter 6 is a very attractive approach to Cornucopia's information needs.

The labor charge overages are troublesome for several reasons. First, they illustrate the difficulty analysts have in developing accurate estimates when it comes to systems work. Second, the number of hours required to develop this project illustrate that although application development software makes the job easier, there is still a considerable amount of work to do. Finally, there is no mention of a discussion between the analysts and the user about the account juggling required to bring the project to completion within the total budget amount.

The above concerns can be addressed one by one. Analysts can improve their estimating abilities if they carefully document their experiences, both good and bad. Analysts can reduce the time required to develop information systems if they take advantage of the productivity enhancing system tools that become available. Analysts can be forthright about estimate errors if they embrace the notion that the analyst and user are partners in these projects.

How to Copy the Concepts

The maintenance and review portion of the SDLC is impossible to simulate in the classroom. The key point to remember is that information systems require regular monitoring in terms of system performance and user needs. In a similar fashion, analysts themselves require regular maintenance. Hopefully, this exercise is only the first of many such episodes.

DISK FILES

There are no files specific to this chapter on the student resource disk.

EXTENDED EXERCISES

LandScapes Extended Exercise 3

a. Review the request for system services (textbook Figure 18-6) and your hardware resource specifications for *LandScapes'* notebook computer (Extended Exercise 1). Evaluate the system services request in terms of your previous hardware specifications. Can the notebook be upgraded to accommodate multimedia? Is $5000 adequate?

b. How would you classify this request? Is this adaptive system maintenance, a system upgrade, or the first signs of functional obsolescence? Draft a response to *LandScapes* describing your findings and recommending a course of action.

Appendix A

Student Resource Disk

Note: The files on the Student Resource Disk were created with soure software as indicated below. The word processed files (.doc and .dok) are exactly the same, except for the font.

File Extension	Source Program
*.doc	WordPerfect 5.1 for Windows, with Century Schoolbook font
*.dok	WordPerfect 5.1 for Windows, with Courier 12pt (10 cpi)
*.bmp	Windows Paintbrush
*.wb1	Quattro Pro for Windows
*.dbf	dBASE IV (DOS)
*.prg	dBASE IV (DOS)

File Name	Description
agenda.doc	This is a team project organization meeting agenda (Figure 1-2).
bill_hrs.doc	This is a generic billable hours form.
budget.wb1	This is the project budget template file.
context.bmp	This skeleton image of Cornucopia's existing system context diagram contains three external entities, each with two data flow lines.
contract.doc	This is the Cornucopia project contract that appears in the textbook.
cost_ben.wb1	This is the cost/benefit template file.
cpia_res.doc	This is the detailed list of resource requirements for the Cornucopia project (textbook Figure 11-9).
cpia_tsk.doc	This document lists all of the detailed tasks for the Cornucopia project.
data_dct.doc	This is the data dictionary entry form.
dfd_lev1.bmp	This skeleton image of Cornucopia's first-level DFD contains four processes, three external entities, four data stores, and several connecting data flow lines.
dfd_sym.bmp	This file contains images used to render the data flow diagrams in the textbook.

erd_sym.bmp This file contains images used to render the entity-relationship diagrams in the textbook.

eval.doc This is the self/peer project team evaluation form (Figure 1-4).

gantt.bmp This is a project management worksheet (Figure 2-5).

guid_wk.bmp This is a blank GUID worksheet.

hardware.doc This is the project hardware resources worksheet (Figure 2-1).

monitor.bmp This is an image of the "Estimated versus Actual" monitoring tables presented in casebook Figure 6-1.

new_usd.bmp This skeleton image of Cornucopia's new system USD contains six subsystems, three external entities, four data stores, and numerous data flows.

old_erd.bmp This skeleton image of Cornucopia's existing entity-relationship diagram contains four entities.

operate.doc This is the project environment operating procedures worksheet (Figure 2-3).

out_wksh.doc This is a blank output worksheet.

pert.bmp This is an image of a generic PERT chart for the major SDLC activities identified in the Cornucopia project.

profile.doc This is the team member profile form (Figure 1-3).

proj_dct.dbf This is an empty dBASE IV file for use as a project dictionary.

review.doc This is a sample project review form.

software.doc This is the project software resources worksheet (Figure 2-1).

status.wb1 This is the project status template file.

struct.bmp This skeleton image of Cornucopia's sales transaction subsystem structure chart contains several functional modules.

sys_sym.bmp This file contains images used to render the system flowcharts in the textbook.

tasks.dbf This is an empty dBASE IV file for recording team member task assignments.

testctl.doc This is the product testing control sheet presented in textbook Figure 15-7.

testproc.doc This is the product testing procedure form presented in textbook Figure 15-7.

weeks.bmp This file contains the project milestone events (Figure 1-1).

The following files implement Cornucopia's database applications.

\mainmenu.prg

\custmain

customer.dbf	**blankcus.prg**
custmain.prg	**custform.prg**
custadd.prg	**browcust.prg**
custchg.prg	**savecust.prg**
custdel.prg	**restcust.prg**

\cdmain

cd.dbf	**blankcd.prg**
cdmain.prg	**cdform.prg**
cdadd.prg	**browcd.prg**
cdchg.prg	**savecd.prg**
cddel.prg	**restcd.prg**

\saletran

sales.dbf	**sales.prg**
dayhist.dbf	**upsales.prg**
saletran.prg	**salesum.prg**
blanksal.prg	

Team Project Organizational Meeting
AGENDA

1. Introductions (9 min.)
    a. Computer skills
    b. Personal schedules and phone numbers
2. Selection of a team meeting facilitator (1 min.)
3. Discussion of team project packet (15 min.)
4. Discussion of team protocol (15 min.)
    a. Team meeting day, time, and place
    b. Agenda format for future meetings
    c. Standards of conduct
5. Distribution of initial task assignments (5 min.)
    a. Task description
    b. Deliverable format and content
    c. Responsible team member or members
    d. Due date
6. Adjourn

Billable Hours Form

Analyst ______________________________

Project	Activity/Task	Date	Hours	Comment

CORNUCOPIA INFORMATION SYSTEM
PROJECT CONTRACT

PROBLEM SUMMARY:

Cornucopia is a small music store specializing in classical records, cassette tapes, compact disks and videos. They want to improve their computer information system in four areas:

1. Customer record keeping
2. Product reordering
3. Customer communications
4. Sales trend analysis

PROJECT SCOPE:

For the present, the new information system will *not* include a computerized cash register and inventory system. Rather, it will introduce a system limited to the capture of customer information and compact disk sales.

The final report will include a preliminary analysis of how the new system could be modified to include "point of sale" technology.

PROJECT CONSTRAINTS:

Cost: The full price of the new system will not exceed $10,000. The product will be under warranty for 30 days from the point of implementation. Thereafter, the on going maintenance costs will not exceed 1% per month of the original system costs.

Delivery Date: The new system will be fully operational within four months of the date of this contract.

Other: Cornucopia will bear the costs of the first 12 hours of personnel training and the cost of the initial customer and CD master file creation. Additional training time will be billed at the rate of $50 per hour.

Objectives: M&M Computer Consultants will deliver a computer information system that provides users with detailed procedures on how to apply hardware and software resources that address the information needs itemized above. In addition, the new system will be designed to:

1. ... <u>not</u> increase the time it takes to complete a normal sale by more than 10%.

2. ... <u>add no more than</u> five hours per week to the time required to perform customer and CD master file maintenance.

3. ... <u>reduce</u> compact disk reordering time by at least 25%.

4. ... <u>increase</u> repeat customer sales by 5% by the end of the first year of operation.

5. ... <u>reduce</u> "out of stock" and "over stock" situations by 10%, over a two year period of time.

Cornucopia - Resource Requirement Specifications

HARDWARE

Processing Platform:

Intel 486DX/33 MHz
64K Cache RAM
4MB RAM
1.2MB & 1.44MB Drives
200MB 13ms IDE Cache Drive
ATI Graphics Ultra Video
14" CrystalScan 1024NI Color VGA Monitor
1 Parallel / 2 Serial Ports
124-Key AnyKey Keyboard
Microsoft Mouse
(bundled software detailed below - #1)
Vendor: Gateway 2000 — $2,395
Source: PC Magazine (9/29/92)

Peripherals:

HP DeskJet 500 — $414
Source: Computer Buying World (9/92)
Gateway 2000 TelePath Fax/Modem — $195
14,400 bps data mode
9,600 bps fax mode
(bundled software detailed below - #2)
Source: PC Magazine (9/29/92)
Uniscan-200 Bar Code Reader — $295
Source: InfoWorld (9/14/92)

SOFTWARE

MS DOS 5.0 & MS Windows 3.1 (#1)
WinFax Pro, Crosstalk for Windows, Qmodem (#2)
Central Point Anti-Virus 1.2 for Windows — $75
Central Point Backup 7.2 for Windows — $75
The Norton Desktop 2.0 and
Prisma Your Way 2.0 for Windows — $95
Vendor: Gateway 2000 (ref. #1 & #2)
Source: PC Magazine (9/29/92)

MS Excel for Windows (#1)0	
WordPerfect for Windows	$260
Borland dBASE IV 1.5	$476
Z-SOFT PC Paintbrush IV+ 1.01	$114

Source: Computer Buying World (9/92)

DATA

None

PEOPLE

Training (bundled with labor cost)

PROCEDURES

Documentation (bundled with labor cost)

MISCELLANEOUS

Paper products (paper, labels)	$100
Floppy disks	$ 50
DeskJet ink supplies	$100
Reference materials	$ 75
Cleaning materials	$ 50
Power Strip Surge Protector	$ 25

SUMMARY

Hardware	$3,399
Software	$1,095
Data	0
People	0
Procedures	0
Miscellaneous	$ 400
	=====
Subtotal	$4,894
Tax & Shipping	$ 500
	=====
Total	$5,394
	******

CORNUCOPIA
DETAILED TASK LIST

ANALYSIS

T.1 - Initial Consultation (4)
- T.1.1 - Interviews (1)
- T.1.2 - Feasibility Report Preparation (2)
- T.1.3 - Contract Preparation (1)

===> E.2 - **Contract Completion**

T.2 - Full Analysis (14)
- T.2.1 - Interviewing (3)
- T.2.2 - Industry Research (2)
- T.2.3 - Existing System Diagramming (4)
  - Context Diagram
  - Data Flow Diagram
  - User's System Diagram
  - Entity-Relationship Diagram
- T.2.4 - Build vs. Buy "Mix" Analysis (1)
- T.2.5 - Develop Project Budget (1)
- T.2.6 - Develop Project Status Report (1)
- T.2.7 - Preliminary Preparation (2)

===> E.3 - **Project Preliminary**

DESIGN & DEVELOPMENT

T.3 - Initial Design Sketch (7)
- T.3.1 - Incorporate Bid Review into Design (1)
- T.3.2 - Develop Alternative New System Proposals (3)
  - User's System Diagram
  - Data Flow Diagram
  - Entity Relationship Diagram
  - Menu Tree
  - Basic I/O Formats
- T.3.3 - Develop Cost/Benefit Analysis (1)
- T.3.4 - Prepare Detailed Hardware & Software Specs. (1)
- T.3.5 - Prepare Design Proposal (1)

===> E.4 - **Design Review**

T.4 - Create Prototype (27)
- T.4.1 - Incorporate Design Review into Design (1)
- T.4.2 - Revise New System Models (2)
- T.4.3 - Design Menu Tree Screens (1)

T.4.4 - Develop Master File I/O's (4)
T.4.5 - Develop Query I/O's (5)
T.4.6 - Develop Report I/O's (5)
T.4.7 - Develop Process Designs (5)
Program Module Structure Charts
System Flowcharts
T.4.8 - "Unit" Testing (2)
T.4.9 - Prepare Prototype Demonstration (2)

===> E.5 - **Prototype Review**

T.5 - Final Product Development (20)

T.5.1 - Incorporate Prototype Review into Design (2)
T.5.2 - 4GL Programming (10)
T.5.3 - Build System Environment (4)
T.5.4 - "System" Testing (4)

===> E.6 - **Training Session**

IMPLEMENTATION

T.6 - Develop System Documentation (8)

T.6.1 - Assemble Project Binder (1)
T.6.2 - Prepare Reference Manual (2)
T.6.3 - Prepare Procedures Manual (5)
System Description
Operating Instructions
User Interface Illustrations
System Security Provisions
Emergency Instructions
Appendices

===> E.7 - **Product Delivery**

T.7 - Develop Training Material (4)

T.7.1 - Establish Training Schedule (1)
T.7.2 - Prepare Training Manual (3)
System Overview
Demonstration Outline
Hands-On Exercises
Quick Reference Guide

===> E.6 - **Training Session**

T.8 - Installation (2)

T.8.1 - Prepare Conversion Plan (1)
T.8.2 - Supervise File Conversion & Creation
(Under our contract, this is primarily a user activity.)
T.8.3 - Project Review (1)
User Acceptance

===> E.7 - **Product Delivery**

Self/Peer Project Team Evaluation

Your Name ____________________ Project ____________

This evaluation is limited to the team project. Complete a section for yourself and each of your team members. Circle the number of points you would award for each performance category.

Self evaluation:

Leadership contribution	1	2	3	4	5
Cooperativeness	1	2	3	4	5
Dependability	1	2	3	4	5
Workload share	1	2	3	4	5
Skills improvement	1	2	3	4	5

Total points ____

Team member ____________________

Leadership contribution	1	2	3	4	5
Cooperativeness	1	2	3	4	5
Dependability	1	2	3	4	5
Workload share	1	2	3	4	5
Skills improvement	1	2	3	4	5

Total points ____

Data Dictionary Entry Form

Element Name: **Type:**

Description:

Contents:

User Cross Reference:

Storage Reference:

Project Hardware Resources

microprocessor ________ memory cache ________

clock speed ___________ RAM ________________

hard disk space ________ disk cache ____________

video bus _____________ monitor ______________

printer ______________ modem ______________

CD-ROM ______________ scanner ______________

____________________ ____________________

____________________ ____________________

____________________ ____________________

____________________ ____________________

Project Environment Operating Procedures

Lab access procedures:

network account name __________ password *******

local account name _____________ password *******

team disk area path ____________________________

lab hours: Mon Tue Wed Thr Fri Sat Sun

------ ----- ------ ----- ----- ----- -------

special privileges or restrictions:

____

____

Software access policy:

site licensed software _____________________

student licensed software __________________

Output Worksheet

Title	________	________	________

Audience

executive	________	________	________
manager	________	________	________
everyday worker	________	________	________

Content

one source	________	________	________
multi-source	________	________	________

Form

hardcopy	________	________	________
softcopy	________	________	________
regularly scheduled	________	________	________
on-demand	________	________	________
report	________	________	________
query	________	________	________
user inquiry	________	________	________

Team Member Profile

Name ________________________ Phone ___________

Availability Schedule:

Time	Mon	Tue	Wed	Thr	Fri	Sat	Sun
------	------	-----	------	------	-----	-----	------
------	------	-----	------	------	-----	-----	------
------	------	-----	------	------	-----	-----	------

Computer Related Course Work and Experience:

Introductory	Applications	Programming
literacy	word processing	Pascal
problem solving	spreadsheet	C
operating systems	database	BASIC
networking	graphics	COBOL
----------------	--------------	-------------
----------------	--------------	-------------
----------------	--------------	-------------
----------------	--------------	-------------

Project Review Form

Evaluate the following items on a scale of 1 to 5.

	Very Poor 1	2	3	4	Very Good 5
1. Problem Summary					
a. Problem statement	1	2	3	4	5
2. Scope					
a. Accuracy of the scope statement	1	2	3	4	5
3. Constraints					
a. Time estimate	1	2	3	4	5
b. Cost estimate	1	2	3	4	5
4. Objectives					
a. Objective 1 product	1	2	3	4	5
b. Objective 2 product	1	2	3	4	5
c.	1	2	3	4	5

Project Software Resources

System software specifications:

network operating system ____________________

peer-to-peer _____ client/server _____

local operating system ____________________

desktop manager software ____________________

utility software ____________________

Application software specifications:

word processing ____________________

spreadsheet ____________________

database ____________________

graphics ____________________

other ____________________

Programming software specifications:

procedural ____________________

nonprocedural ____________________

Product Testing Control Sheet

Project: ________________

Category and Product Description	Comment

Syntax
*
*
*
*

User Interface
*
*
*
*

Integrated Module
*
*
*
*

System
*
*
*
*

Pilot
*
*
*
*

User Acceptance
*
*
*
*

Product Testing Procedure

Project: ________________ **Product:** ____________________

Test Description: ____

Test Data **Expected Results** **Comment**

*
*
*
*

*
*
*
*

*
*
*
*

*
*
*
*

*
*
*
*

*
*
*
*
*
*

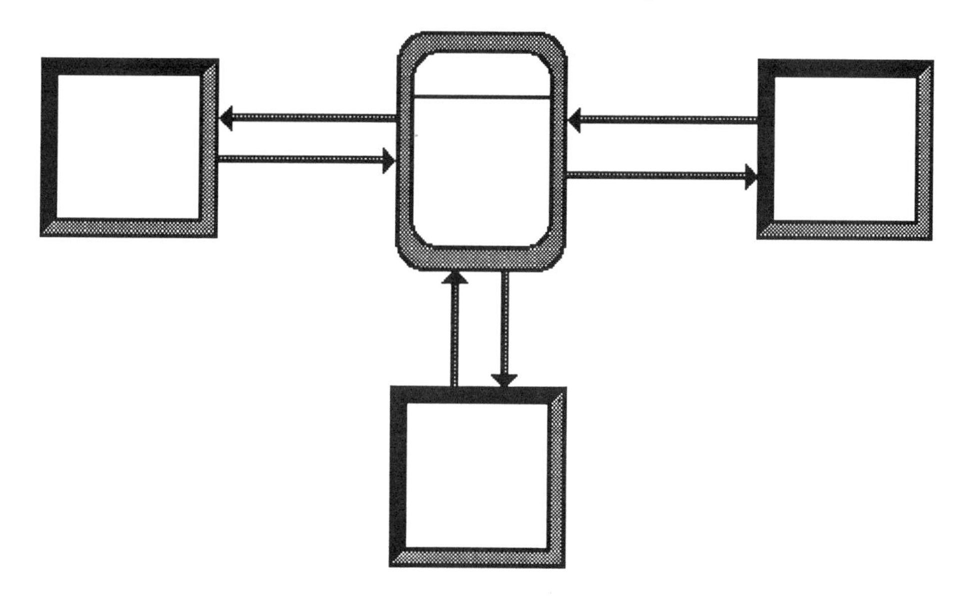
Skeleton System Context Diagram

Skeleton First-Level DFD

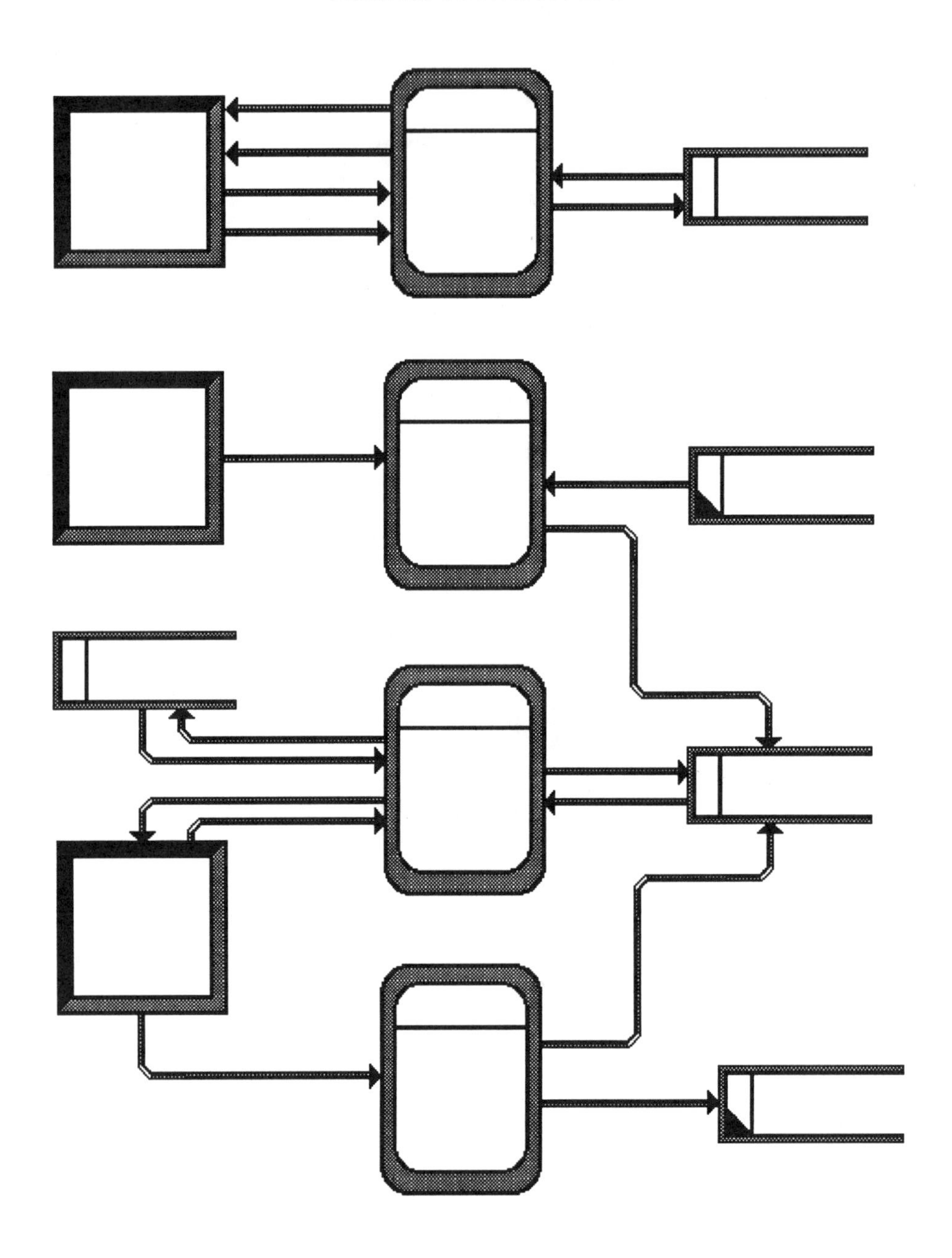

Data Flow Diagram Symbols

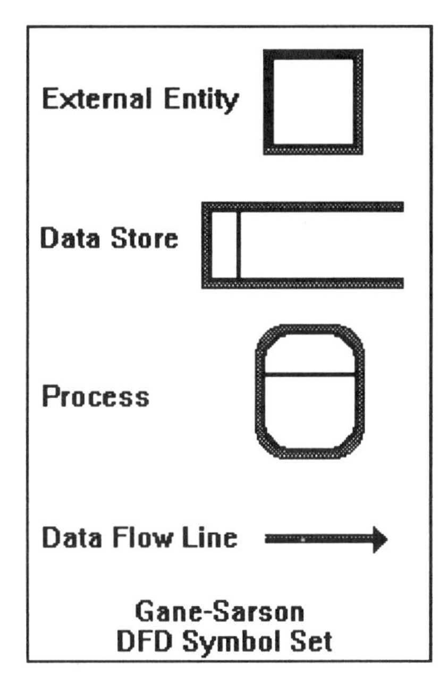

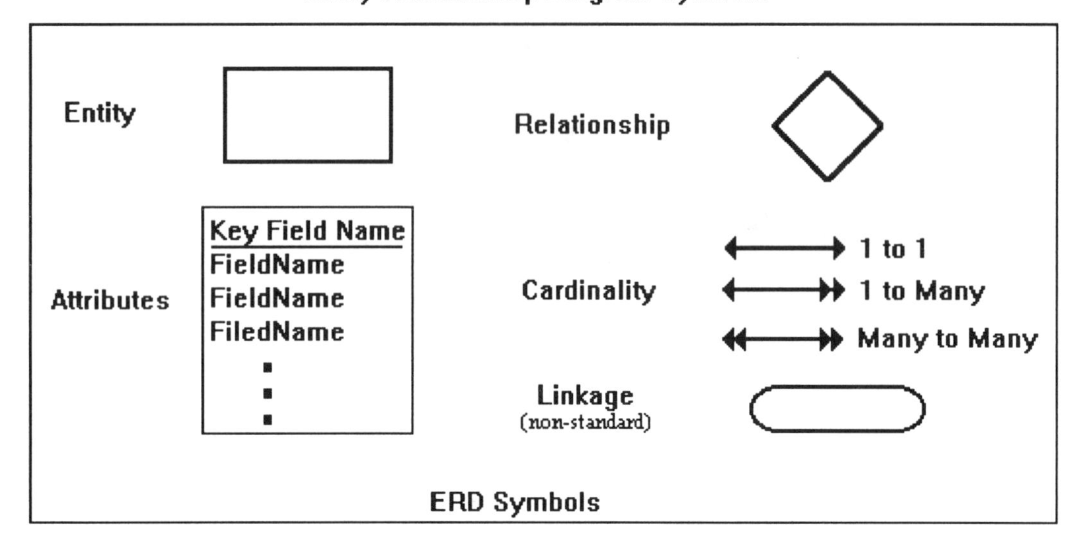
Entity-Relationship Diagram Symbols
Entity
Relationship
Attributes
Key Field Name
FieldName
FieldName
FiledName
Cardinality
1 to 1
1 to Many
Many to Many
Linkage
(non-standard)
ERD Symbols

gantt.bmp

Project Management Work Sheet

As of:

			Time Period																
			1	2	3	4	5	6	7	8	9	10	11	12	13	14	15	16	Total

monitor.bmp

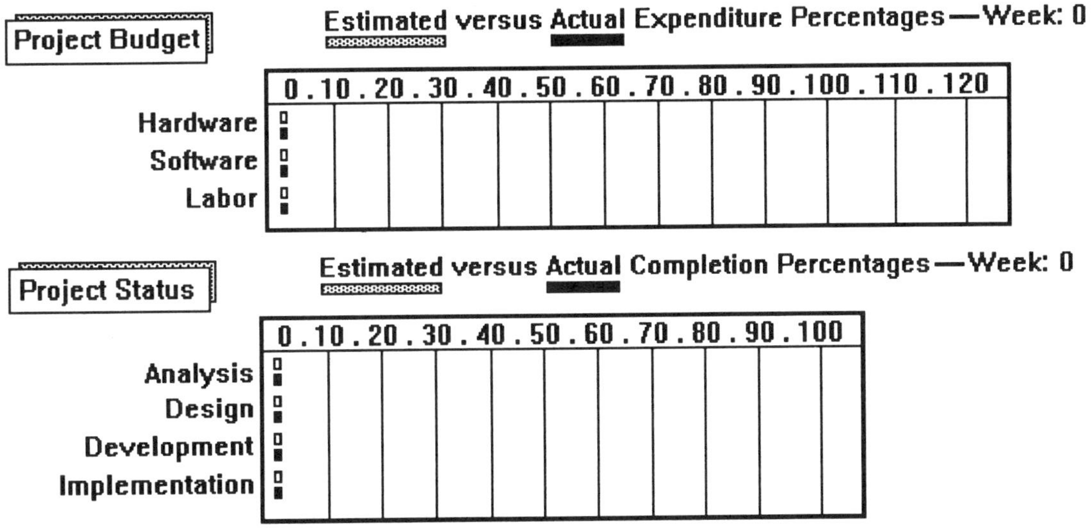
Project Budget
Estimated versus Actual Expenditure Percentages—Week: 0
0 . 10 . 20 . 30 . 40 . 50 . 60 . 70 . 80 . 90 . 100 . 110 . 120
Hardware
Software
Labor
Project Status
Estimated versus Actual Completion Percentages—Week: 0
0 . 10 . 20 . 30 . 40 . 50 . 60 . 70 . 80 . 90 . 100
Analysis
Design
Development
Implementation

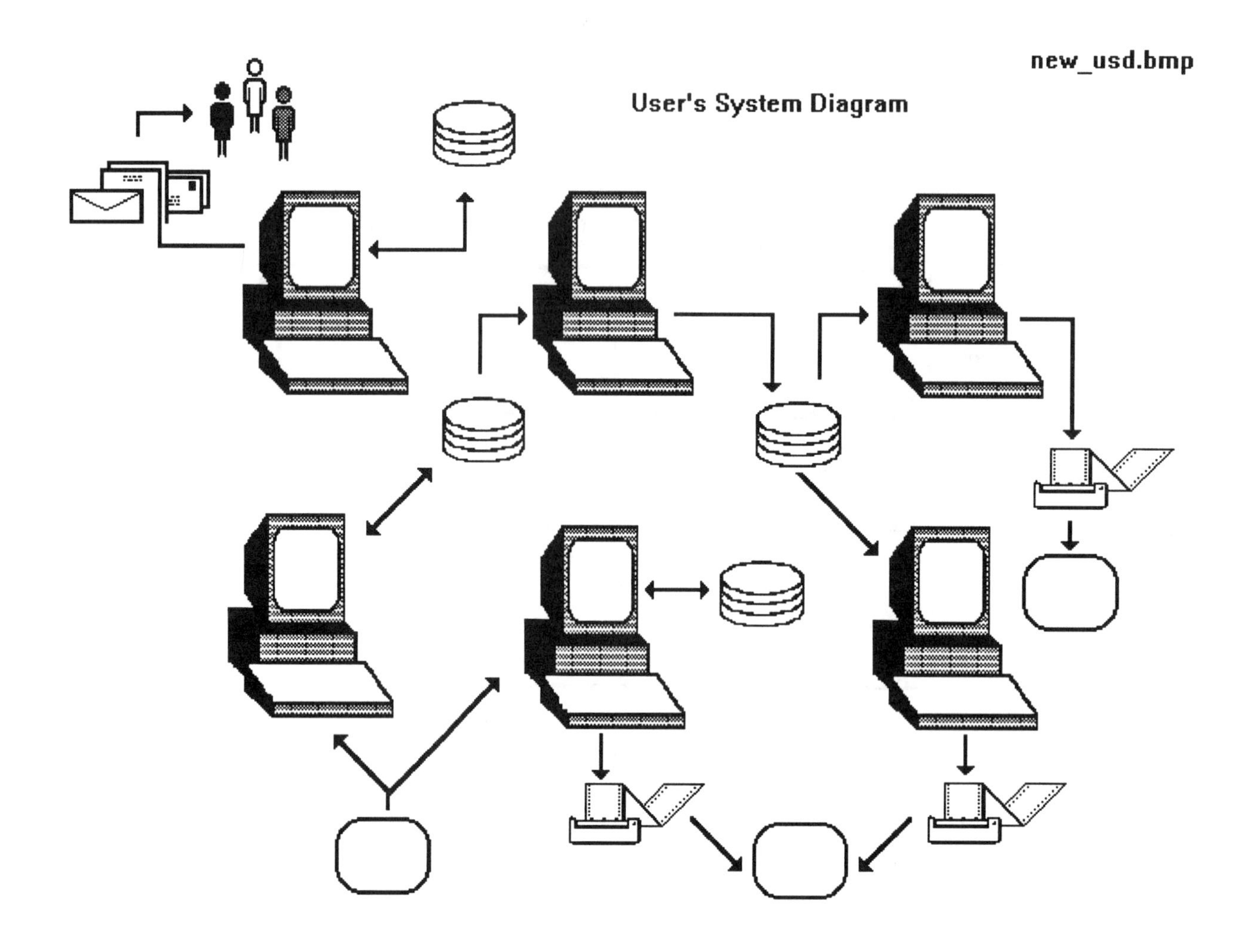
new_usd.bmp
User's System Diagram

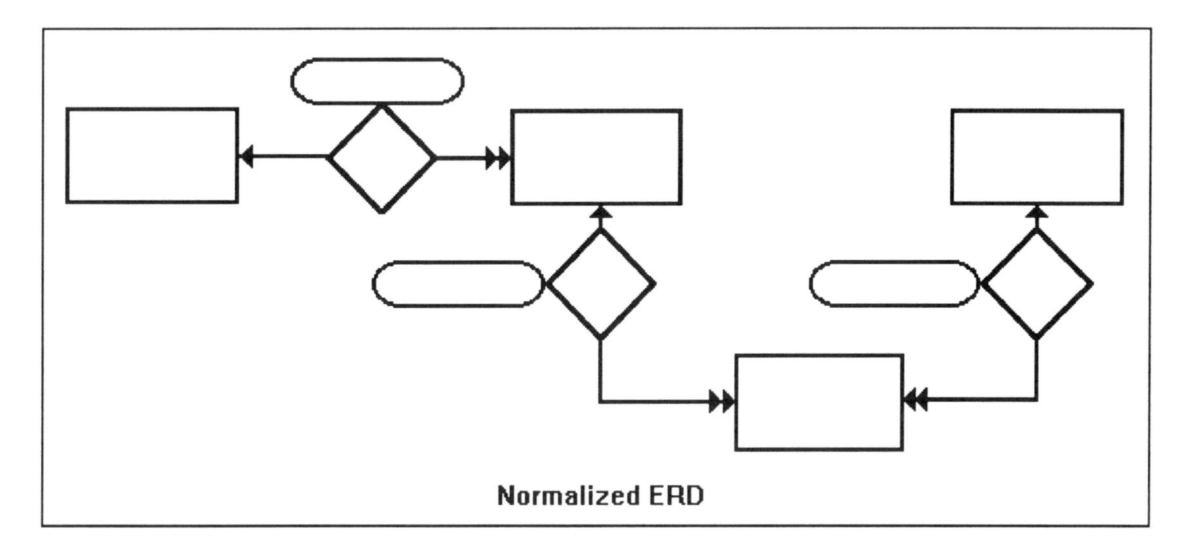
Normalized ERD

Cornucopia PERT Chart

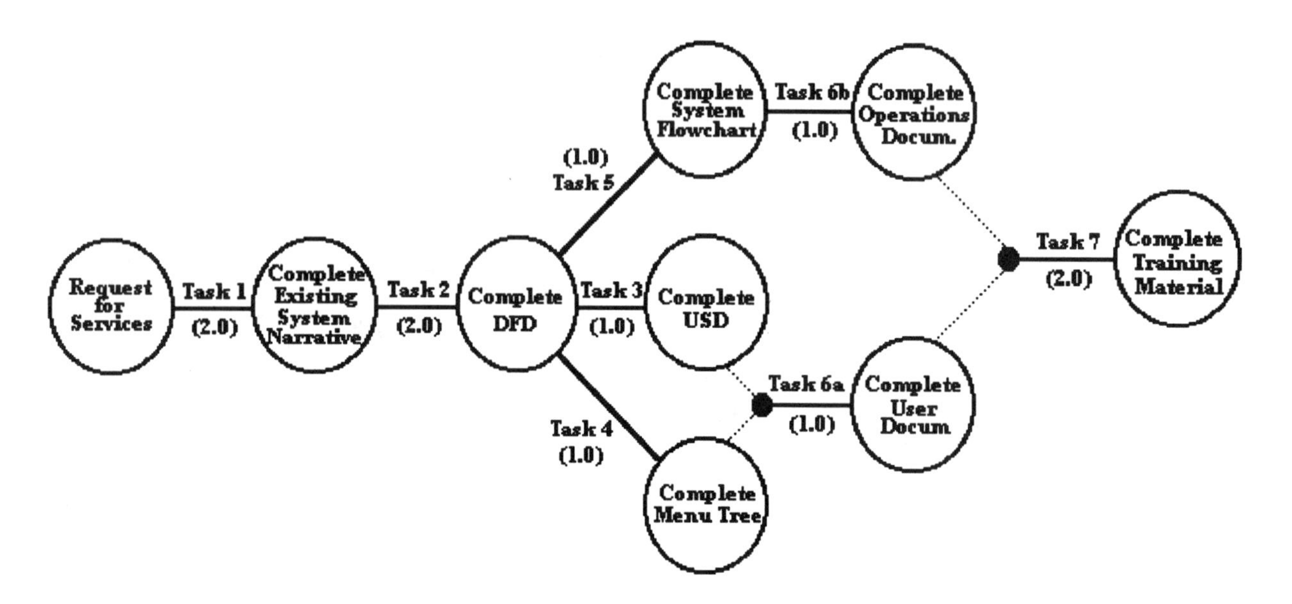

Subsystem Structure Chart

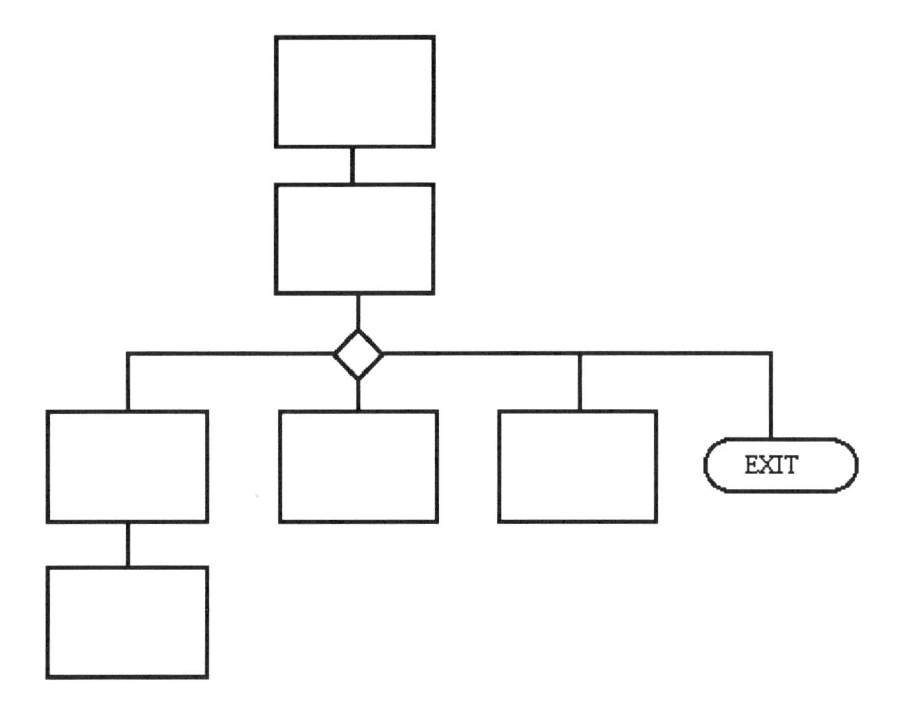

System Flowchart Symbols

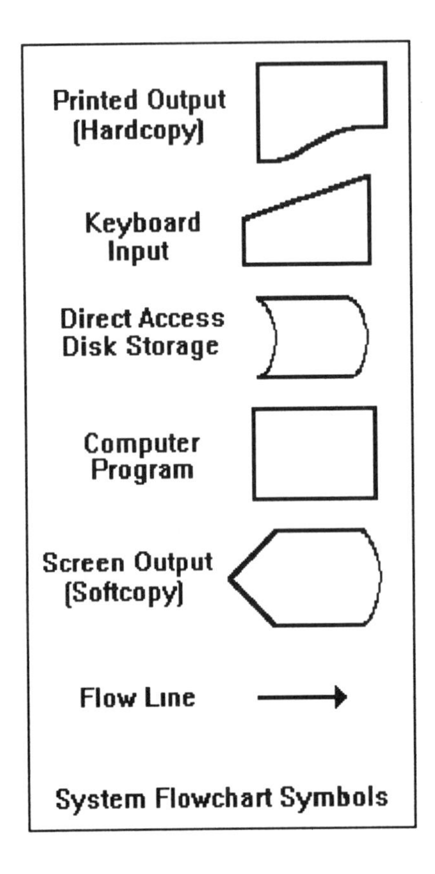

weeks.bmp

Project Milestone Events

Week 1	Week 2	Week 3	Week 4	Week 5	Week 6	Week 7	Week 8	Week 9	Week 10	Week 11	Week 12	Week 13	Week 14	Week 15	Week 16
Project Contract		Project Preliminary Session					Design Review Session			Prototype Review Session				Training Session	Final Report

Week 1	Week 2	Week 3	Week 4	Week 5	Week 6	Week 7	Week 8	Week 9	Week 10	Week 11	Week 12	Week 13	Week 14	Week 15	Week 16

budget.wb1

Date: Budget A/O Wk: 1

	Week1	Week2	Week3	Week4	Week5	Week6	Week7	Week8	Week9	Week10	Week11	Week12	Week13	Week14	Week15	Week16	Total
Estimates																	
Hardware																	0
Software																	0
Labor																	0
Total	0	0	0	0	0	0	0	0	0	0	0	0	0	0	0	0	0
Actuals																	
Hardware																	0
Software																	0
Labor																	0
Total	0	0	0	0	0	0	0	0	0	0	0	0	0	0	0	0	0
Weekly +/-																	
Hardware	0	0	0	0	0	0	0	0	0	0	0	0	0	0	0	0	0
Software	0	0	0	0	0	0	0	0	0	0	0	0	0	0	0	0	0
Labor	0	0	0	0	0	0	0	0	0	0	0	0	0	0	0	0	0
Total	0	0	0	0	0	0	0	0	0	0	0	0	0	0	0	0	0
Cumm. +/-																	
Hardware	0	0	0	0	0	0	0	0	0	0	0	0	0	0	0	0	
Software	0	0	0	0	0	0	0	0	0	0	0	0	0	0	0	0	
Labor	0	0	0	0	0	0	0	0	0	0	0	0	0	0	0	0	
Total	0	0	0	0	0	0	0	0	0	0	0	0	0	0	0	0	

cost_ben.wb1

Date: Cost/Benefit Analysis

Month =>	1	2	3	4	5	6	7	8	9
Costs:									
Initial Sys.									
Training									
Maintenance									
Other Tangible									
Intangible									
Benefits:									
Tangible									
Intangible									
================	========	========	========	========	========	========	========	========	========
Cummulative Costs	0	0	0	0	0	0	0	0	0
Cummulative Benefits	0	0	0	0	0	0	0	0	0

status.wb1

Date: Project Status A/O Week: 1

Activity	% Comp.	Status	1	2	3	4	5	6	7	8	9	0	1	2	3	4	5	6	Total
Analysis-Est																			0
Actuals																			0
Design -Est																			0
Actuals																			0
Develop-Est																			0
Actuals																			0
Impl. -Est																			0
Actuals																			0
Total -Est			0	0	0	0	0	0	0	0	0	0	0	0	0	0	0	0	0
Actuals			0	0	0	0	0	0	0	0	0	0	0	0	0	0	0	0	0
Contract																			
Prelim																			
Design Rev.																			
Proto. Rev.																			
Train & Del.																			
Final Rpt.																			
			1	2	3	4	5	6	7	8	9	0	1	2	3	4	5	6	

Appendix B

The Cornucopia Case

Chapter 3
Problem Identification and Definition

Each step of the SDLC is illustrated by the Cornucopia Case, which appears in a series at the end of the remaining chapters. Based on an actual small-enterprise case, the series reinforces the concepts presented in each chapter and serves as a model for your own parallel case study.

Background

Cornucopia is a small music store located in the "Old Town" section of the business district. Its marketing "signature" is the personal attention it provides its patrons. Coupled with an extensive classical musical inventory of compact disks, cassettes, records, and videos is a modest collection of books on classical music. The customers can take pleasure in a wealth of knowledge about the composers, performing artists, and particular recordings by engaging either the owner or one of the two sales clerks, who seem to always make time to discuss what is of obvious interest to them. All this commerce and conversation takes place amid a few potted plants, beautiful classical background music, and complimentary apple juice or wine.

The last thing the enterprise needs is an expensive, impersonal, complicated computer information system. Yet, the following letter describes some problems that the owner would like to solve (Figure 3-8).

Notice that the owner did not prepare a request for system services. This is not unusual. Small-enterprise users are not bound by bureaucratic procedures. Indeed, they relish their independence. We will simply work around this shortcoming by conducting a careful preliminary investigation and preparing a good project contract.

After two meetings with the owner, during which we were able to observe the business operation first-hand, we confirm that the owner's original letter is accurate in its description of the present system. We also determine that the owner is indeed serious about this project and is willing to invest in a modest feasibility study. The cost of the study will either be absorbed into the project contract or priced not to exceed $200 if the project is aborted.

Feasibility Analysis

The first challenge to the analyst is to determine, in the most general sense, whether he or she can offer a solution to the explicit problems the owner has identified. (Note: At this point, none of the information needs are unusual or pose particularly difficult technical problems. These conditions are by design, so that we can concentrate on the teaching points of the SDLC. Other complications may arise in the future, but for now, it is assumed that a popular microcomputer platform, along with several off-the-shelf software packages, will provide the raw resources for the solution).

Cornucopia Initial Problem Statement

FIGURE 3–8

"Without music, life would be a mistake."

January 1, 1995

M & M Computer Consultants
1130 Avenue of the Giants
Myers Flat, CA 95554

Dear Sir or Madam:

You were referred to me by one of your former clients, who recommended that my business could benefit from a computer information system. Let me tell you a little about how we operate and what we have in mind.

To begin with, we would like to improve our customer record keeping procedures. All transactions involve cash, personal check, or credit card. Thus, this procedure does not involve accounts receivable. However, we do keep a partial file of customer names and addresses on an Apple II computer, which is located at my home. This information is taken from personal checks or at the request of customers who wish to be placed on the mailing list. It is used to create mailing labels for periodic promotional material.

Second, we would like to improve our reordering system. At present it relies entirely on the observations of the employees. When they notice a particular shortage, they make note of it on a spiral pad next to the cash register. Periodically, I place orders, via telephone, to the appropriate wholesaler and note the date of the order on the pad. Special orders for disks, cassettes, videos, and books follow the same procedure, except that they are recorded on separate, color-coded sheets of paper. As shipments are received, these notes are scratched off.

Third, we would like to develop a quarterly newsletter for our customers. This would contain information on subjects of interest to our customers, as well as provide advertising for the enterprise.

Fourth, we would like to have information about the sales trends of the business. For example, what products generate the most sales, the most profits, etc.

I hope that you can help us. Please call me at 445-9876 so that we can arrange a meeting to discuss this project in more detail.

Sincerely,

Margaret Height

A combination of word processing, spreadsheet, and database software can be tailored to meet all the information needs itemized so far. A rough budget projects $4,000 for hardware purchases, $2,000 for software purchases, and $4,000 for our labor (80 hours at $50 per hour). This budget reflects a major premise of this text, that is, microcomputer-based information systems, relying on well-integrated horizontal software, are well within the economic reach of the small enterprise. It also provides a point of reference for future comparison, as a detailed budget is prepared and actual expenses begin to accumulate.

The next challenge to the analyst is to determine the broad constraints of the project. How much time and money is the owner willing to devote to the project? Are those constraints realistic? Is the owner flexible about considering modifications in either the constraints or the project objectives? The analyst's experience will generally guide these negotiations, unless the owner has predetermined, rigid budgets and time frames in mind. The analyst must remember that his or her enthusiasm about systems work, it is in everyone's interest to discontinue the project if all parties cannot agree on what is feasible and practical.

The Cornucopia owner agrees to our rough budget estimate on two conditions: The project must be completed within four months, and the system must return financial rewards equal to or exceeding the cost within three or four years. We can agree on these terms for the following reasons.

First, our estimate of 80 total labor hours assumes that those hours will be spread out over many weeks, with many interruptions and other small projects intervening from time to time. Four months provides adequate flexibility on this point. Second, our preliminary estimate is that the new system will not add a significant amount of work to the current staff duties. Further, we expect that improved customer communications, combined with fewer lost sales due to out-of-stock problems, will increase sales. A modest estimate of only two additional sales per day, at a net profit of $5 per sale, generates $2,600 per year (260 days x $10). At this rate, a $10,000 system is paid for in a little less than four years.

The Feasibility Report

It is essential that all this analysis be documented, even if only for the analyst's personal reference. The feasibility report is the term that describes this collection of notes, samples, and documents. It can be very formal, or it can be nothing more than a bulging manila folder. For Cornucopia, we choose to prepare a formal report for submission to the owner. This accomplishes two things: It reassures the owner that the project contract to follow is well supported by careful analysis; also, it adds an important element of discipline by forcing the analyst to organize, summarize, and analyze the findings, even for projects that never make it beyond this stage. Figure 3-9 provides a "snapshot" of this report.

Project Contract

We can now develop a written statement that defines the problem and the constraints as agreed to by the analyst and the owner. Although there is no formal request for system services in this case, the original letter from Cornucopia and the feasibility report provide ample information with which construct the project contract (Figure 3-10). Notice that some items in this contract are new to our discussion—a brief explanation follows, with more explanation in later chapters.

FIGURE 3–9 **Cornucopia Feasibility Report**

M&M COMPUTER CONSULTANTS

January 15, 1995 Page 1 of 3

Feasibility Report for Cornucopia

This document summarizes the efforts to date concerning the Cornucopia project (Ref. M.Height, Jan, 1, 1995).

The problems outlined in the original letter are well articulated. We found no extenuating circumstances that would significantly alter the four specific problem areas identified, namely:

1. Customer Master File Maintenance
2. Product Reordering

M&M COMPUTER CONSULTANTS

January 15, 1995 Page 2 of 3

M&M COMPUTER CONSULTANTS

January 15, 1995 Page 3 of 3

We conclude that this project can be completed within the constraints outlined above and that sufficient economic benefits will accrue over the life of the system to justify the cost of the system.

If this project is approved, we will prepare a contract reflecting the analysis contained in this report.

The computerized cash register and point-of-sale issues raised in the scope section reflect the owner's desire to minimize the disruptions in the store due to the new computer system. We agree to this, but anticipate that as the employees and customers become accustomed to the new system, the owner will want to expand it to handle the complete sales transaction. An automated inventory system is another obvious addition, which we should consider when we specify the hardware requirements later.

Information system maintenance and review is a concern to both the user and the analyst. The user does not want to be without some protection against system malfunction. Nor does any user want to go to the Yellow Pages and start all over with a new analyst every time the system might need a slight improvement. At the same time, the analyst wants to develop a good reputation. As any successful entrepreneur will confirm, the after-sale service is just as important as the sale itself. Therefore, we agree to provide regular maintenance service beyond the standard 30-day product warranty period. Extraordinary services, such as a major system upgrade or the initiation of a new SDLC, would require a new contract.

User training is discussed in detail in Chapter 16, but suffice it to say that 12 hours of personnel training is included in the price of the Cornucopia system. The user manuals and system documentation should provide adequate reference beyond that time. Naturally, some ongoing training and user assistance will occur during the regular monthly maintenance visits.

Finally, the objectives section of the contract itemizes several measurable goals while promising to deliver a system that addresses all five components of a computer information system. The specificity of these goals may seem highly speculative. However, they are based on a combination of the analyst's experience, observation of the enterprise operations, and discussions with the owner. As such, they represent an honest attempt to quantify the performance standards for the project.

Cornucopia Information System Project Contract

FIGURE 3–10

CORNUCOPIA INFORMATION SYSTEM
PROJECT CONTRACT

Problem Summary:

Cornucopia is a small music store specializing in classical records, cassette tapes, compact disks, and videos. They want to improve their computer information system in four areas:

1. Customer record keeping
2. Product reordering
3. Customer communications
4. Sales trend analysis

Project Scope:

For the present, the new information system will not include a computerized cash register and inventory system. Rather, it will introduce a system limited to the capture of customer information and compact disk sales.

The final report will include a preliminary analysis of how the new system could be modified to include "point of sale" technology.

Project Constraints:

Cost: The full price of the new system will not exceed $10,000. The product will be under warranty for 30 days from the point of implementation. Thereafter, the ongoing maintenance costs will not exceed 1 percent per month of the original system costs.

Delivery Date: The new system will be fully operational within four months of the date of this contract.

Other: Cornucopia will bear the costs of the first 12 hours of personnel training and the cost of the initial customer and CD master file creation. Additional training time will be billed at the rate of $50 per hour.

FIGURE 3–10 **Cornucopia Information System Project Contract**

CORNUCOPIA INFORMATION SYSTEM
PROJECT CONTRACT

Page 2 of 2

Objectives:

M&M computer Consultants will deliver a computer information system that provides users with detailed procedures on how to apply hardware and software resources that address the information needs itemized above. In addition, the new system will be designed to:

1. ... not increase the time it takes to complete a normal sale by more than 10 pecent.

2. ... add no more than five hours per week to the time required to perform customer and CD master file maintenance.

3. ... reduce compact disk reordering time by at least 25 percent.

4. ... increase repeat customer sales by 5 percent by the end of the first year of operation.

5. ... reduce "out of stock" and "over stock" situations by 10 percent over a two-year period of time.

Chapter 4
Data Flow Diagrams

The analysis phase for the project is now well underway. The analyst must first understand the existing system before the new system can be designed. The process model is the first abstraction to develop.

The Context Diagram

The owner's original letter, the fact-finding activities, and the subsequent discussions involving the project contract provide the information necessary to prepare the context diagrams for the existing system (Figure 4-13). Notice that the existing system does not include the owner as an external entity because she never originates input or receives output. All her activities are internal to the system. From the project contract we know that the new system, on the other hand, will produce at least two new information products: newsletter and sales trend reports. Therefore, the owner, in receiving the sales trend reports, becomes an external entity to the new system. Also notice that the sales system is an external entity because it is a separate information system within the enterprise.

The First-Level DFD

The next step in the analysis process is to develop the first level DFD of the existing system. Once again, the previous activities provided the insight necessary to identify the specific tasks and task IPO charts that will help us decompose the context diagram. Of the four areas in this project, only the "reordering system" and the "customer record-keeping procedures" currently exist. The narrative below is based on the owner's original letter.

FIGURE 4–13 Cornucopia Existing System Context Diagram

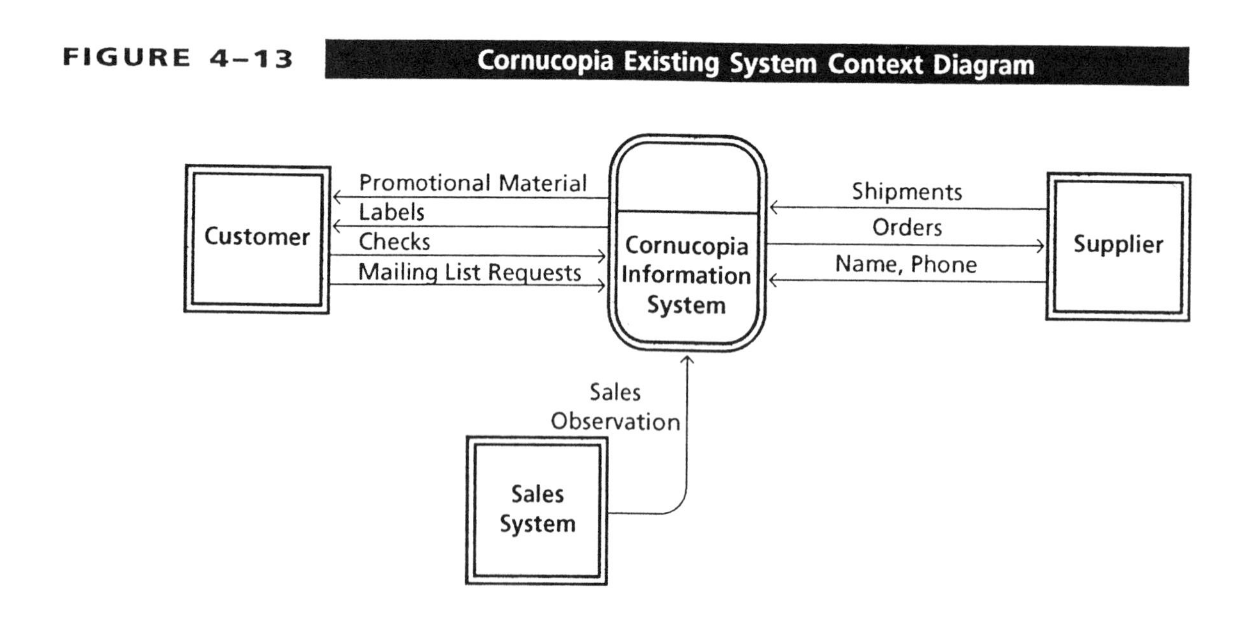

The customer record-keeping procedure does not involve accounts receivable, because all transactions involve cash, check or credit card. A partial file of customer names and addresses is kept on an Apple II computer at the owner's home. The owner updates this file from customers' personal checks or direct requests from the customer. Mailing labels are printed from this system. The computer is also used to create the promotional materials that are periodically mailed to customers.

The reordering system relies entirely on employees' observations. By observing sales activity and/or the inventory directly, they notice stock shortages, which they make note of on a spiral pad next to the cash register. Periodically, the owner consults a suppliers list and then places orders, via telephone, to the appropriate wholesaler and finally, notes the date of the order on the pad. Special orders for disks, cassettes, videos and books follow the same procedure, except that they are recorded on separate, color-coded sheets of paper. As shipments are received, these notes are scratched off.

Several action words or phrases are underlined in the narrative, from which four cohesive tasks emerge, as detailed in the following outline. Note that the customer record-keeping task is simply called customer correspondence and exists completely separately from the other three tasks. In other words, the current system provides no cross-reference between customers and their specific purchases. This fact was confirmed through a last-minute phone call to the owner.

TASK: Customer Correspondence
- INPUTS:
  - personal checks (Source: Customer)
  - customer requests (Source: Customer)
  - customer name, address (Data Store Retrieval: Customer)
- OUTPUTS:
  - customer name, address (Data Store Update: Customer)
  - mailing labels (Sink: Customer)
  - promotional material (Sink: Customer)

TASK: Inventory
- INPUTS:
  - observations of inventory (Data Store Retrieval: Inventory)
  - observations of sales (Source: Sales System)
- OUTPUTS:
  - notes the shortage (Data Store Update: Orders)

TASK: Order
- INPUTS:
  - notes the order (Data Store Retrieval: Orders)
  - consults supplier list (Data Store Retrieval: Supplier)
  - supplier changes (Source: Supplier)
- OUTPUTS:
  - places orders (Sink: Supplier)
  - notes the date (Data Store Update: Orders)
  - supplier changes (Data Store Update: Supplier)

TASK: Receive
  INPUTS:
    shipments (Source: Supplier)
  OUTPUTS:
    scratch off order (Data Store Update: Orders)
    shipments (Data Store Update: Inventory)

Figure 4-14 presents the first-level DFD of the existing system. Although this diagram could be decomposed further, doing so would not add significantly to our understanding of the system. One interesting observation about this process is that not only does it help us to understand the current system, but it also stimulates some early thinking about the design of the new system. For example, the four data stores are possible database candidates. Also, as previously noted, the current reordering system does not interact with the customers or the sales transactions. However, such interaction is likely in the new system, given the requirement to add customer communications and sales trend analysis.

Time and Money

The Cornucopia project is already accumulating costs, consuming analyst time, and making some progress toward meeting its goals. Although project management is not discussed until Chapter 6, it is never too soon to mention that the project-budget and completion-status reports are two important tools used to monitor such data. So, until Chapter 6, let's simply note that the analyst has spent a total of four hours on this project, for a cost of $200 ($50 per hour). Further, the analyst estimates that about 30 percent of the analysis phase is complete.

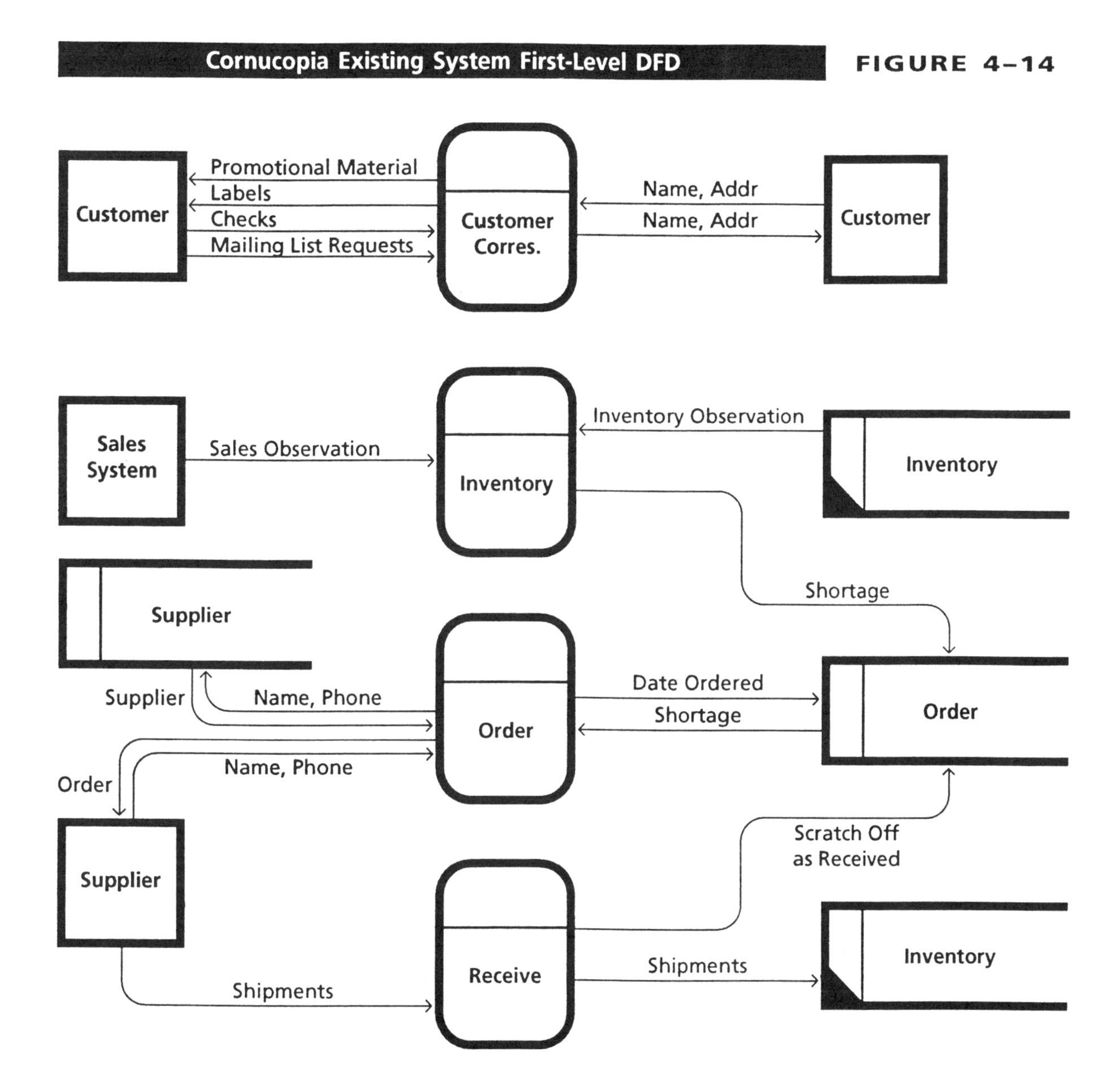
Cornucopia Existing System First-Level DFD
FIGURE 4–14
Customer
Promotional Material
Labels
Checks
Mailing List Requests
Customer Corres.
Name, Addr
Name, Addr
Customer
Sales System
Sales Observation
Inventory
Inventory Observation
Inventory
Shortage
Supplier
Supplier
Name, Phone
Order
Date Ordered
Shortage
Order
Order
Name, Phone
Supplier
Scratch Off as Received
Receive
Shipments
Inventory
Shipments

Chapter 5
System and Data Models

The analysis of the current system continues with the development of system models, namely the user's system diagram. The data model for Cornucopia's existing system not only proves to be interesting but also provides some new insights into the project.

The System Model

The first-level DFD developed in Chapter 4 (Figure 4-13) can be transformed into the more user-friendly USD, which substitutes familiar icons for the formal symbols and eliminates some of the detail. The first challenge for the analyst is to choose appropriate icons. However, clip art and paintbrush packages make this task fairly easy. The next challenge for the analyst is to label each icon. The analyst should use verbs or verb phrases to label processes. This will make it easier to distinguish processes from external entities and data stores, which are labeled with nouns or noun phrases. Figure 5-15 shows the USD for the current reordering system.

FIGURE 5–15 Cornucopia Existing System USD

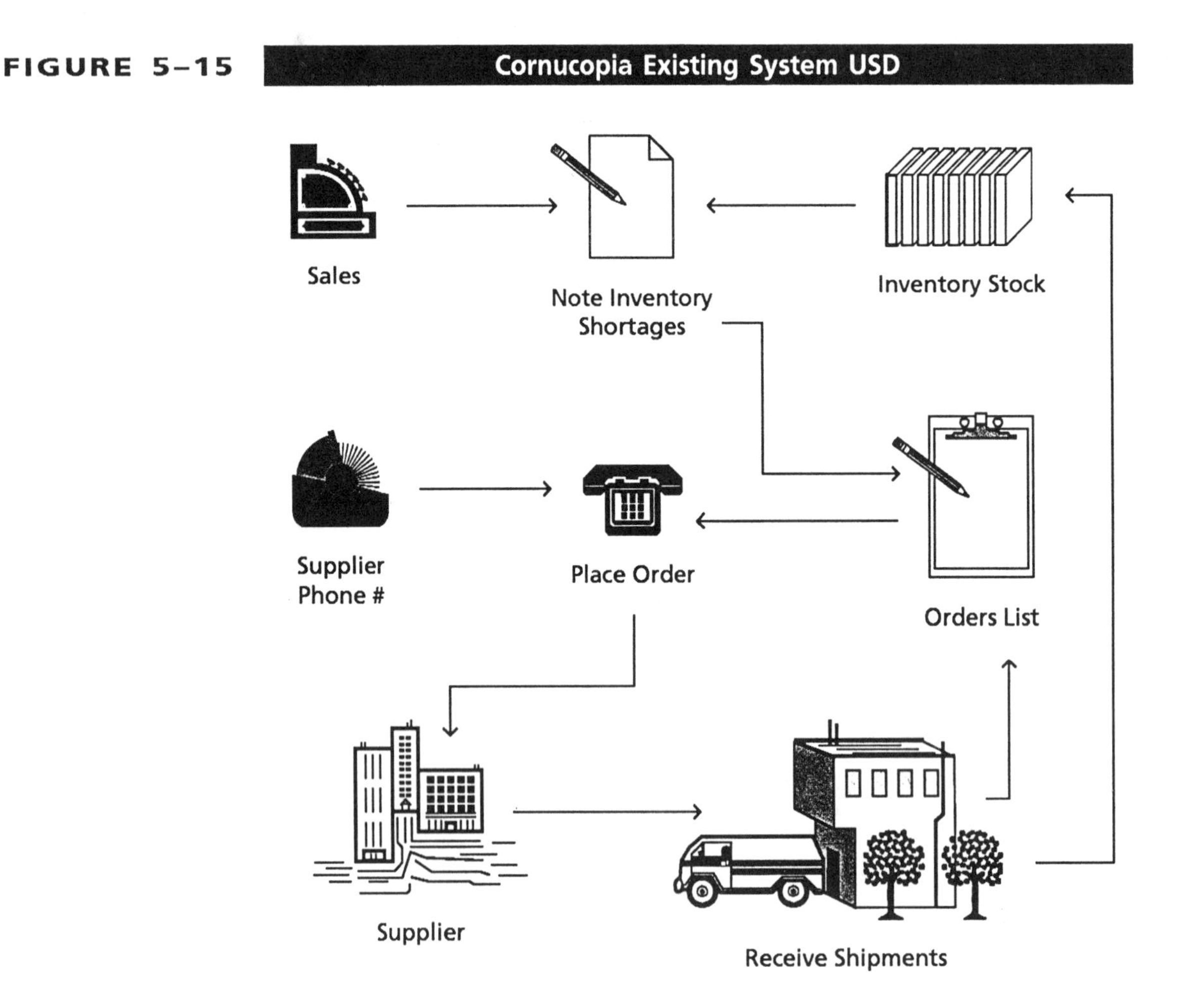

Remember that the purpose of this model is to facilitate communication between the analyst and the user. A meeting with the owner confirms that this indeed is the way the reordering system works today. She is intrigued by the artwork, but she wonders how necessary it is and how much it cost to prepare. When we pull out the first-level DFD and explain the importance of understanding the current system in order to avoid costly redesigns later, she is temporarily satisfied. She shows no interest in discussing the DFD.

Because the current system is entirely manual, we did not need to prepare a menu tree diagram or a system flowchart. This saves time and avoids further explanations to the owner about our methodology. Also notice: We did not include customer correspondence on this USD. For now, this element of the system is totally independent of the reordering system and therefore would unnecessarily clutter the diagram.

The Data Model

Even though "customer" is the only existing system data store that is computerized, we can still use the data modeling process to further our understanding of the current system. We begin with the data flow diagram, which identifies four data stores: customer, supplier, order, and inventory. Our goal is to prepare an entity-relationship diagram using these four data stores as the entities. Each entity is defined by its name, its attributes, its relationship to the other entities, and its cardinality within these relationships.

The attributes for each entity are detailed in the following table. This information was gathered partly through the feasibility study and partly through a series of follow-up question-and-answer sessions with the owner.

customer =	supplier =	order =	inventory =
name +	name +	order# +	UPC +
address +	phone#	date +	title +
phone#		UPC +	artist +
		amount +	label +
		received	price

We can make several interesting observations about these entities and their attributes. First, you must realize that the owner does not have a piece of paper that documents this information. This business enterprise is not concerned about such detailed computer jargon. Only after much coaching do we learn that the owner simply "knows" which supplier to call for which products. Likewise, the order# attribute is definitely not subject to a systematic, sequential numbering procedure. The order# is nothing more that a penciled digit on the top of the page. Each day begins a new sequence, starting at 1. This explains why both order# and date are required to form the key attribute for this file. The UPC (for "universal product code") is the only attribute that appears in more than one file.

The cardinality relationship between supplier and order is one-to-many. That is, for every order there is only one supplier, because no one wants to order the same disk or tape from two suppliers at the same time. What is very likely, however, is that over time several orders will be placed to the same supplier, resulting in that supplier's name or number appearing more than once in the order file.

The order and inventory file presents a many-to-many cardinality relationship. A single UPC may appear more than once in the order file and a single order# may contain more than one UPC. Accordingly, we need to normalize the file relationship by creating an order/inventory file with the following attributes:

order/inventory =
order# +
date +
UPC

Figure 5-16 presents the entity-relationship diagram for the existing system. Notice that the customer entity does not interact with any of the other entities, which is consistent with the information on the data flow diagram.

FIGURE 5–16 Cornucopia Existing System ERD

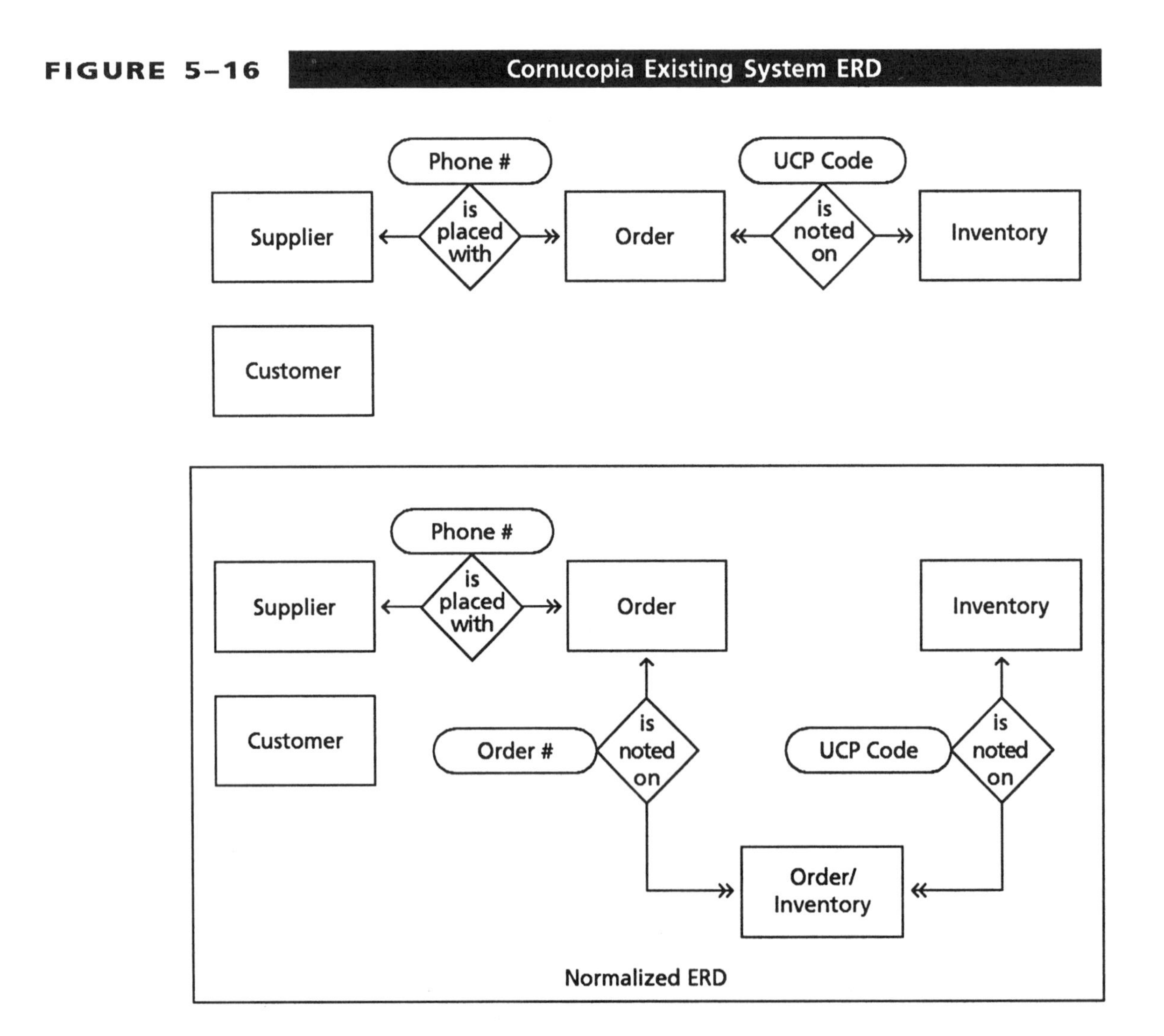

Collectively, we now have several views of the existing system, including the narration, the DFD, the USD, and the ERD. Using the USD to "set the tone," we have established a working relationship with the user that will continue throughout the project.

Ideas for the New System

Our efforts to date have not focused totally on the old system. The project contract specifies four areas for improvement:

Areas for Improvement

1. Customer record keeping
2. Product reordering
3. Customer communications
4. Sales trend analysis

Each of the analysis phase activities stimulate ideas for the new system as outlined in these broad terms, but two thoughts are very clear at the moment. First, the current system modeling confirms that no provision exists for capturing sales data for future sales trend analysis. This suggests that the design will need to incorporate a sales subsystem of some kind. Second, the reordering system relies far too much on the owner's memory. The new design will need to provide some consistent way to associate supplier phone numbers and the products that Cornucopia needs to order.

The analyst should make notes of such design ideas as they occur. Not only does this provide a stimulus to the formal design process, but it also offers an important insight into the complexity of the project, which will certainly help the analyst develop a more detailed project budget and project status report.

Time and Money

The analyst spent five hours developing the system and data models for the existing system. This time was evenly split between desk work and field research—the analysis phase is not a solitary activity. It actively involves the analyst and the owner. Remember,enterprise personnel are prime sources for information about how the existing system works.

The analyst estimates that the analysis phase is now about 75 percent complete. In addition, one of the five hours was devoted to some preliminary sketches of the new system design, which allows the analyst to report a 5 percent completion of the design phase.

Time "billed" to any given project is not necessarily expended in one time frame. Many unrelated interruptions and intervening activities are inevitable. Furthermore, it is common for an analyst to work on more than one project at a time. While waiting for an answer on one project, the analyst can work on another project. This demands that the analyst develop a procedure to record and report project hours accurately—and to the correct project.

Chapter 6
Project Management

At this point in the process we have a good understanding of the old system and a rough idea of what the new system will include. The owner is aware of our general approach to the project, but we have not formally presented our ideas. To structure our efforts and assure the owner that we have the necessary project controls in place, we must prepare several project management products and present a coherent plan. The Project Preliminary will include a written and oral presentation to the owner. The following items are to be included in this presentation:

Written Material:
1. Summary of Project Requirements
2. Overview of General Design Concepts
3. Proposed Timetable (Project Status Report)
4. Proposed Project Budget

Oral Presentation:
1. Introductions
2. Visuals for Items 1-4 Above
3. Question and Answer Period
4. Preview of Next Step

"Build versus Buy"

Before we proceed much farther, however, we must consider the results of our industry and vertical software research. As mentioned in Chapter 3, we don't want to "reinvent the wheel." Complete turnkey products may be available for less money than the one we propose to assemble. A little research shows that an existing point-of-sale (POS) product is available, one that can considerably reduce our work: Ozware Computer Systems markets several variations of "Phono-Scan," at an initial cost of between $1,990 and $4,990, plus $700 for installation, another $1,000 for peripheral equipment, then $1,280 in annual update and maintenance charges. The full-featured system comes to $6,690, plus $1,280 annual fees. The system does not include the customer correspondence subsystem elements outlined by the user, but we could add this feature and still keep the overall cost below $10,000.

Although tempted, we won't adopt this vertical product in our case study. After all, we still have 12 chapters and several weeks to go before the end of the academic term. It is instructive, however, to see that real-world products do indeed that serve small-enterprise computing needs.

The Project Budget

Budgets can be built from the "bottom up" or from the "top down." Mostly they are built with some of each method. This project has some very clear constraints: don't spend more than $10,000, and complete it in 16 weeks. The three major cost components are: hardware, software, and labor. With these givens, and a little experience, the budget is constructed using the format established earlier in the chapter. Figure 6-11 presents the budget as of Week 3.

This budget reflects an estimate of 86 labor hours to complete the project. The hardware and software purchases are estimated to occur during weeks 6 through 9. This budget update shows that we have not billed as many labor hours as estimated, which is shown as a $250 balance in the cumulative section.

FIGURE 6–11 **The Cornucopia Project Budget**

Date: Cornucopia Budget A/O Wk: 3

	Week1	Week2	Week3	Week4	Week5	Week6	Week7	Week8	Week9	Week10	Week11	Week12	Week13	Week14	Week15	Week16	Total
Estimates																	
Hardware						2500	500	500	500								4000
Software						1000	250	250									1500
Labor	200	250	250	400	200	250	250	400	200	250	250	400	250	250	250	250	4300
Total	200	250	250	400	200	3750	1000	1150	700	250	250	400	250	250	250	250	9800
Actuals																	
Hardware																	0
Software																	0
Labor	100	200	150														450
Total	100	200	150	0	0	0	0	0	0	0	0	0	0	0	0	0	450
Monthly +/–																	
Hardware	0	0	0	0	0	0	0	0	0	0	0	0	0	0	0	0	0
Software	0	0	0	0	0	0	0	0	0	0	0	0	0	0	0	0	0
Labor	100	50	100	0	0	0	0	0	0	0	0	0	0	0	0	0	250
Total	100	50	100	0	0	0	0	0	0	0	0	0	0	0	0	0	250
Cum. +/–																	
Hardware	0	0	0	0	0	0	0	0	0	0	0	0	0	0	0	0	
Software	0	0	0	0	0	0	0	0	0	0	0	0	0	0	0	0	
Labor	100	150	250	0	0	0	0	0	0	0	0	0	0	0	0	0	
Total	100	150	250	0	0	0	0	0	0	0	0	0	0	0	0	0	

The Project Status Report

Figure 6-12 presents the project status report as of Week 3. Notice that the 86 labor hours are distributed across the phases of the project as follows:

Project Phase	Estimated Hours	Percentage
Analysis	18 hours	20.9%
Design	27 hours	31.4%
Development	27 hours	31.4%
Implementation	14 hours	16.3%

We have billed 9 hours to this project so far, with the analysis phase 75 percent complete and the design phase 5 percent complete. The analysis phase completion percentage coincides with our estimated completion figure, but we accomplished this in 6 fewer hours than we estimated. Therefore, we show a "+ +" in the status column to indicate that we are performing better than planned. Work on the design phase actually began a week earlier than expected, which accounts for the difference in the estimated and actual completion percentage. The contract and preliminary presentation are completed.

Cornucopia Project Status—Week 3

FIGURE 6–12

Date: **Cornucopia Project Status** **A/O Week: 3**

Activity	% Comp.	Status	1	2	3	4	5	6	7	8	9	0	1	2	3	4	5	6	Total
Analysis -Est	75%		4	5	5	4													18
Actuals	75%	+ +	2	4	2														8
Design -Est	0%					4	4	5	5	4									22
Actuals	5%	ok			1														1
Develop -Est	0%									4	4	5	5	4					22
Actuals	0%	ok																	0
Impl. -Est	0%													4	5	5	5	5	24
Actuals	0%	ok																	0
Total -Est			4	5	5	8	4	5	5	8	4	5	5	8	5	5	5	5	86
Actuals			2	4	3	0	0	0	0	0	0	0	0	0	0	0	0	0	9
Contract	100%	ok	C																
Prelim. Pres.	100%	ok			C														
Design Rev.	0%	ok								S									
Proto. Rev.	0%	ok											S						
Train & Del.	0%	ok															S		
Final Rpt.	0%	ok																S	
			1	2	3	4	5	6	7	8	9	0	1	2	3	4	5	6	

The Cornucopia Detail Task List

The following details the Cornucopia tasks, subtasks and events. They correspond to the activities suggested by the SDLC model first presented in Chapter 2. Future chapters describe each of these activities in detail.

Notice that the design and development stages are combined to reflect the back-and-forth relationship between design and prototyping activities. The numbers following each task coincide with the job estimates in the status and budget reports.

Cornucopia
Detailed Tasks

Analysis:

T.1 - Initial Consultation (4)
- T.1.1 - Interviews (1)
- T.1.2 - Feasibility Report Preparation (2)
- T.1.3 - Contract Preparation (1)

===> E.2 - **Contract Completion**

T.2 - Full Analysis (14)
- T.2.1 - Interviewing (3)
- T.2.2 - Industry Research (2)
- T.2.3 - Existing System Diagramming (4)
  - Context Diagram
  - Data Flow Diagram
  - User's System Diagram
  - Entity-Relationship Diagram
- T.2.4 - Build versus Buy "Mix" Analysis (1)
- T.2.5 - Develop Project Budget (1)
- T.2.6 - Develop Project Status Report (1)
- T.2.7 - Preliminary Preparation (2)

===> E.3 - **Preliminary Review**

Design and Development:

T.3 - Initial Design Sketch (7)
- T.3.1 - Incorporate Preliminary Review into Design (1)
- T.3.2 - Develop Alternative New System Proposals (3)
  - User's System Diagram
    - Data Flow Diagram
    - Entity Relationship Diagram
    - Menu Tree
    - Basic I/O Formats
- T.3.3 - Develop Cost/Benefit Analysis (1)
- T.3.4 - Prepare Detailed Hardware and Software Specs. (1)
- T.3.5 - Prepare Design Proposal (1)

===> E.4 - **Design Review**

T.4 - Create Prototype (27)
- T.4.1 - Incorporate Design Review into Design (1)
- T.4.2 - Revise New System Models (2)
- T.4.3 - Design Menu Tree Screens (1)
- T.4.4 - Develop Master File I/O's (4)
- T.4.5 - Develop Query I/O's (5)
- T.4.6 - Develop Report I/O's (5)
- T.4.7 - Develop Process Designs (5)
  - Program Module Structure Charts
  - System Flowcharts
- T.4.8 - "Unit" Testing (2)
- T.4.9 - Prepare Prototype Demonstration (2)

===> E.5 - **Prototype Review**

T.5 - Final Product Development (20)
- T.5.1 - Incorporate Prototype Review into Design (2)
- T.5.2 - 4GL Programming (10)
- T.5.3 - Build System Environment (4)
- T.5.4 - "System" Testing (4)

===> E.6 - **Training Session**

Implementation:

T.6 - Develop System Documentation (8)
- T.6.1 - Assemble Project Binder (1)
- T.6.2 - Prepare Reference Manual (2)
- T.6.3 - Prepare Procedures Manual (5)
  - System Description
  - Operating Instructions
  - User Interface Illustrations
  - System Security Provisions
  - Emergency Instructions
  - Appendices

===> E.7 - **Product Delivery**

T.7 - Develop Training Materials (4)
- T.7.1 - Establish Training Schedule (1)
- T.7.2 - Prepare Training Manual (3)
  - System Overview
  - Demonstration Outline
  - Hands-On Exercises
  - Quick Reference Guide

===> E.6 - **Training Session**

T.8 - Installation (2)
- T.8.1 - Prepare Conversion Plan (1)
- T.8.2 - Supervise File Conversion and Creation
  - (Under our contract, this is primarily a user activity.)
- T.8.3 - Project Review (1)
  - User Acceptance

===> E.7 - **Product Delivery**

The PERT Chart

The detailed task list is used to create the following PERT worksheets. The PERT chart illustrated in Figure 6-13 is derived from this detailed information. The chart shows that tasks T.6 (Develop System Documentation) and T.7 (Develop Training Material) can be worked on at the same time as several other tasks. The critical path for this PERT chart is T.1 through T.5 and T.8.

Cornucopia
PERT Worksheet

Major Events	Week Due
E.1 - Start	Week 1
E.2 - Contract Completed	Week 1
E.3 - Bid Review	Week 3
E.4 - Design Review	Week 8
E.5 - Prototype Review	Week 11
E.6 - Training Session	Week 15
E.7 - Product Delivered	Week 16

Major Tasks	Labor Estimate
Analysis:	
T.1 - Initial Consultation	4 hours
T.2 - Full Analysis	14 hours
Design and Development:	
T.3 - Initial Design Sketch	7 hours
T.4 - Create Prototype	27 hours
T.5 - Final Product Development	20 hours
Implementation:	
T.6 - Develop System Documentation	8 hours
T.7 - Develop Training Materials	4 hours
T.8 - Installation	2 hours

Event/Task Relationship

Event	Begins Task	Ends Task
E.1	T.1	---
E.2	T.2	T.1
E.3	T.3	T.2
E.4	T.4,T.6	T.3
E.5	T.5,T.7	T.4
E.6	T.8	T.5,T.7
E.7	---	T.8,T.6

FIGURE 6–13 **Cornucopia PERT Chart**

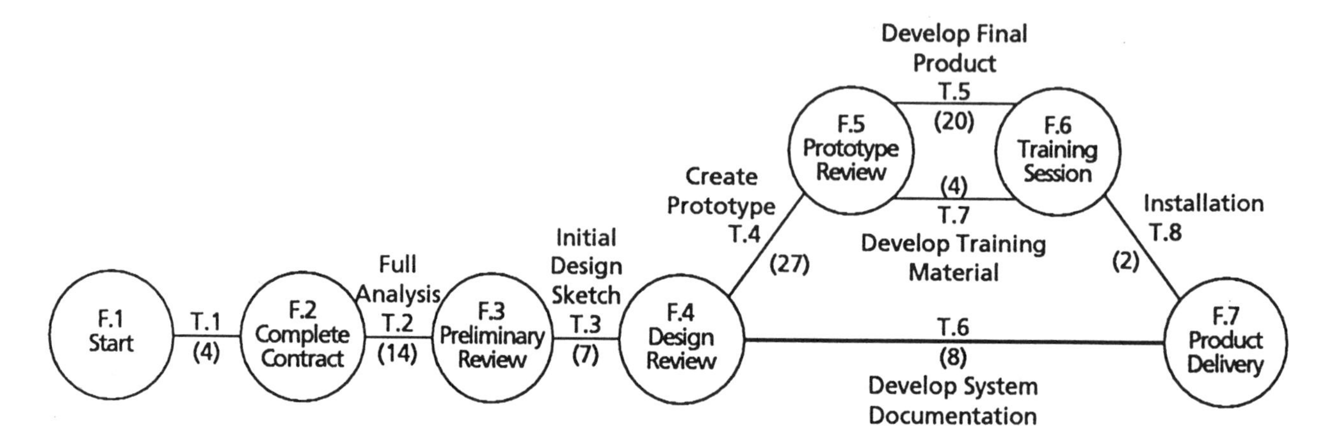

Chapter 7
Problem Solving and Design Development Strategies

As we embark on the design activities of this project, our first task is to develop a preliminary view of the new system design. As noted in Chapter 5, the project contract and our analysis phase activities present a fairly clear picture of what the new system must provide. The following summarizes these requirements:

Areas for Improvement

1. Customer recordkeeping
2. Product reordering
3. Customer communications
4. Sales trend analysis

Ideas for the New Design

1. Incorporate a sales subsystem
2. Associate suppliers with products

The New System USD

You will remember that the existing system USD (Figure 5-15) shows only the reordering system. Our first thoughts about design for the new system are also centered on the reordering system. The owner has already shown us what is important here. She wants a series of sales trend reports to better manage her inventory dollars and improve sales. This can be done only if we capture the sales transaction data in a Sales subsystem and incorporate a Sales Trend subsystem. CD and Supplier subsystems seem reasonable, so that we can create reports that identify specific product trends and orders and provide supplier telephone numbers. In a similar fashion, the request for customer record-keeping and communications leads us to pencil in a Customer subsystem. Counting the obvious Reordering subsystem, we have six processes to work with, as follows:

New Subsystems

1. Sales (sales transaction capture)
2. Sales Trends (sales summarization)
3. Reordering (CD orders based on sales)
4. Customer (customer master file)
5. CD (CD master file)
6. Supplier (supplier master file)

Understand that this is a very preliminary look at the new system. But, it does give us the chance to develop a new system USD (Figure 7-14) that we can share with the owner. Notice that this diagram builds on that of the existing system. This allows the owner to place the new system elements in a familiar frame of reference. The owner's first reaction is, "I thought we were talking about one computer, not six!" Of course, only one computer will be used, but it will feature several menu selections that correspond to each subsystem. We have chosen the computer icon to alert the owner that this is a machine operation.

FIGURE 7–14

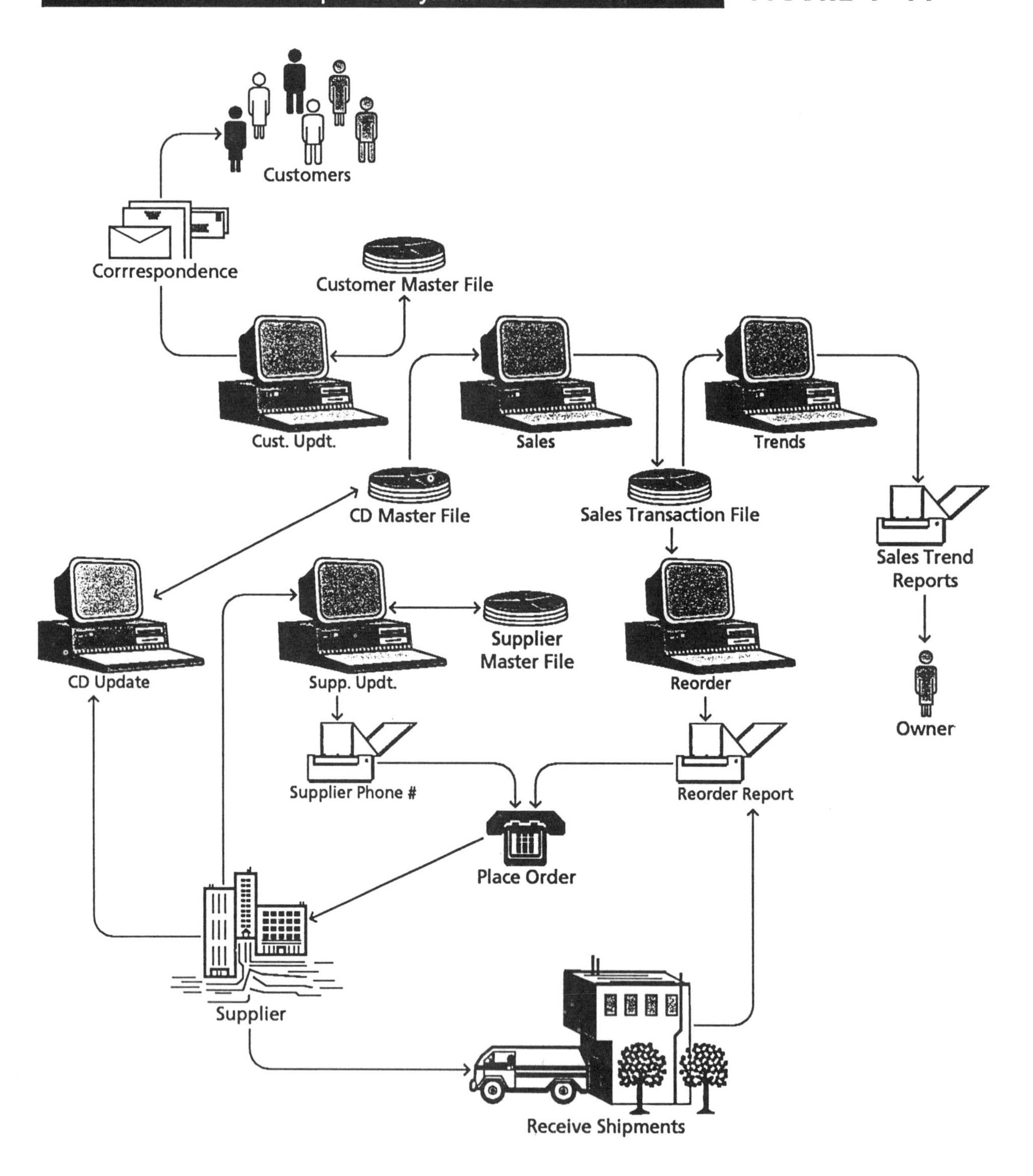

Cornucopia New System USD
Customers
Corrrespondence
Customer Master File
Cust. Updt.
Sales
Trends
CD Master File
Sales Transaction File
Sales Trend
Reports
Owner
CD Update
Supplier
Master File
Supp. Updt.
Reorder
Supplier Phone #
Reorder Report
Place Order
Supplier
Receive Shipments

Time and Money

During the fourth week of the project we spent 3 hours completing the analysis phase and 3 hours on the design phase, as reflected on the Billable Hours forms that appear in Figure 7-15. The analysis phase is virtually 100 percent complete, but because we understand that the design work that lies ahead may require us to reopen some part of the analysis, we record this as only 95 percent complete. With the first attempt at the new system USD completed, we show the design work as 20 percent complete. The updated project status report and budget are omitted.

FIGURE 7–15 Billable Hours Forms—Week 4

M&M COMPUTER CONSULTANTS — Billable Hours Form

Analyst: Jamie

Project	Activity/Task	Date	Hours	Comment
corncpia	T.2.3	2/22	.5	Existing USD
corncpia	T.2.7	2/24	1.0	Finish Prelim. Pres.
corncpia	T.3.1	2/24	1.0	Prelim. Pres. Review

M&M COMPUTER CONSULTANTS — Billable Hours Form

Analyst: Scott

Project	Activity/Task	Date	Hours	Comment
corncpia	T.2.2	2/23	1.5	Industry Res.
silhouet	T.1.2	2/24	1.5	Feasibility Rpt.
corncpia	T.3.2	2/25	2.0	New USD

Chapter 8
File and Form Design

You should note that as we move more into the user-centered activities of the SDLC, we shall highlight the owner's role as a principal user of the system by referring to her as *the user*.

The new USD (from Figure 7-14) provides the springboard for the first serious discussions with the user about the new system design. Our first JAD working session is focused on the USD. The user's initial concern about the multiple computer icons is satisfied after we explain that they represent menu options, not separate computers. We also discuss the consequences of the fact that the Customer Maintenance and Correspondence subsystem is not connected to the rest of the information system. The user agrees that this is something that we should look at as a future system enhancement. Finally, the user insists that the proposed Supplier Master File Update subsystem is unnecessary at this time. She agrees to annotate the supplier phone number listing with information concerning the product lines each supplier handles.

The New System DFD

The analyst, comfortable with the knowledge that the user is in basic agreement with the new system design, returns to model building. First the process model (DFD) is constructed from the USD, then the data model (ERD) is constructed from the previous two models. This exercise serves the same purpose as it did during the analysis of the existing system. It helps the analyst better understand the relationships among the external entities, processes, data flows, and data stores in the new system. Also, on a very practical level, it helps to identify the file types (master, transaction, history) and file servicing forms that must be designed and prototyped during the next JAD working session.

The new data flow diagram (Figure 8-12) differs only slightly from the new USD. The Correspondence subsystem and the Form Letters data store are added to fully describe the plan to improve the customer recordkeeping and communications portion of the information system. During our JAD working session to discuss the general system design, the user indicated that she wants to receive a copy of the reorder report. As mentioned, during this session the user also expressed her desire to eliminate the Supplier subsystem. Nevertheless, you can see that the Supplier Update subsystem is still on the DFD. This is because the subsystem remains a part of the information system even though it is a manual operation. Remember, these models are abstractions and do not necessarily define a specific implementation medium for the data stores or process.

The New System ERD

The data model, shown as the ERD in Figure 8-13, is derived from the data flow diagram. The data stores on the DFD become the data entities on the entity-relationship diagram. Notice that the relationship file Customer/Letters has been added to eliminate the many-to-many relationship between customers and form letters. Also notice that the paper-based supplier master file is included in the diagram.

These two model-building activities do not involve the user in the same way as does the building and verification of the user's system diagram. The user will be brought back into the design process during the form design and prototyping work that is soon to follow.

Cornucopia New System First-Level DFD (Version 1.0) FIGURE 8–12

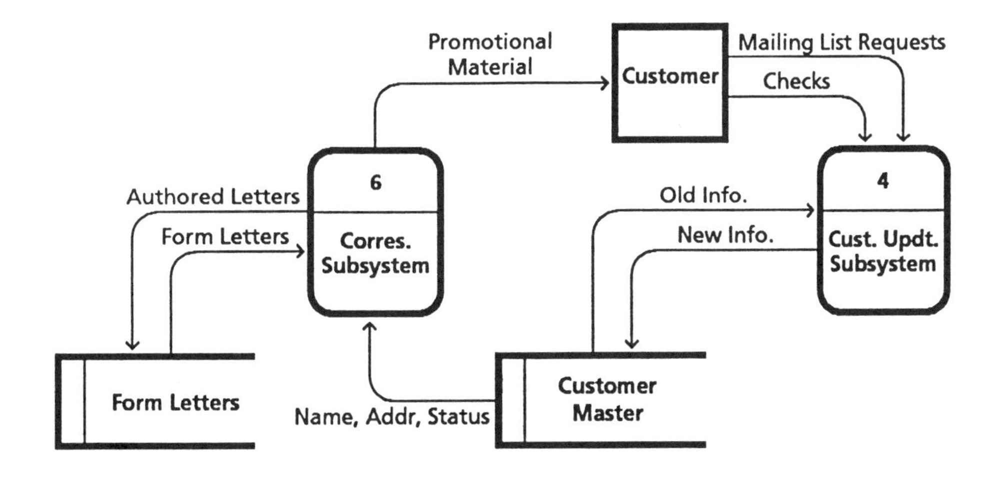

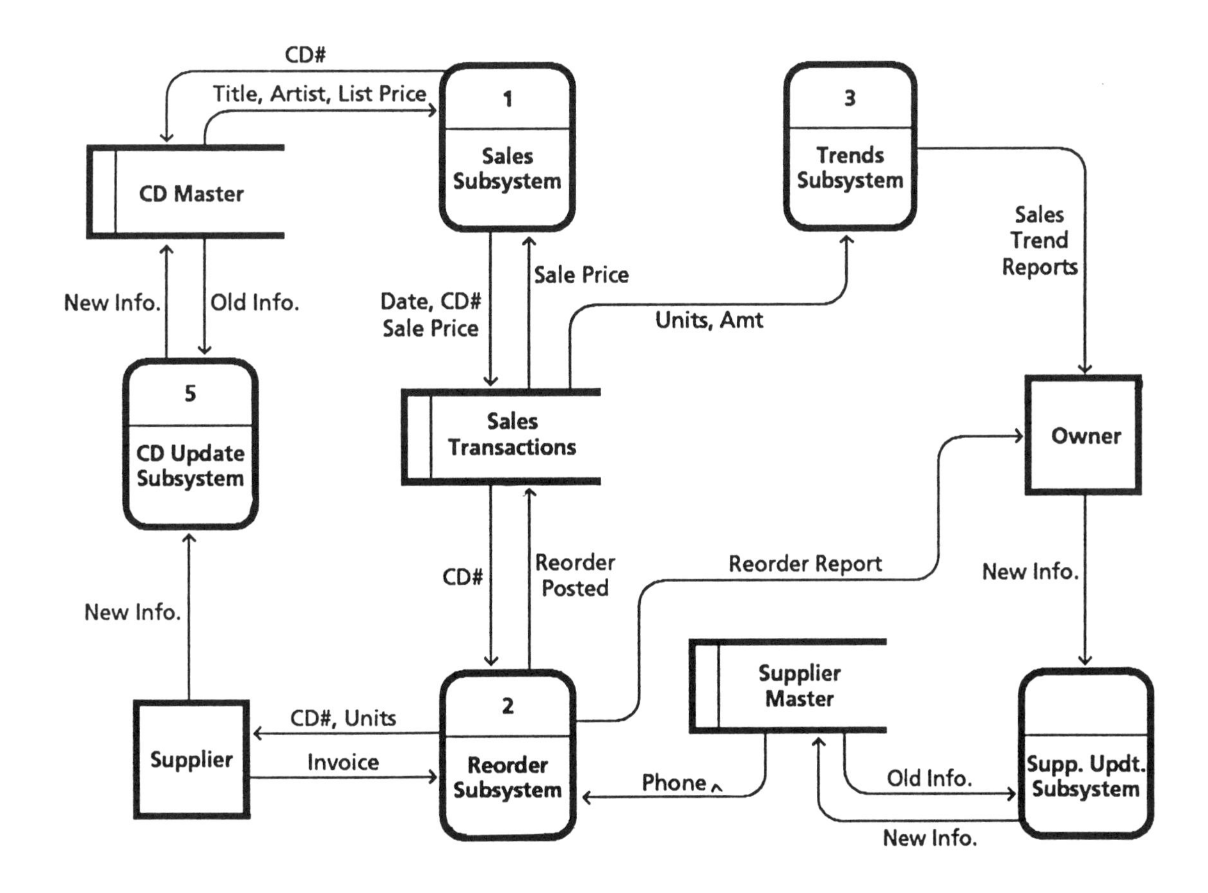

FIGURE 8–13 Cornucopia New System ERD (Version 1.0)

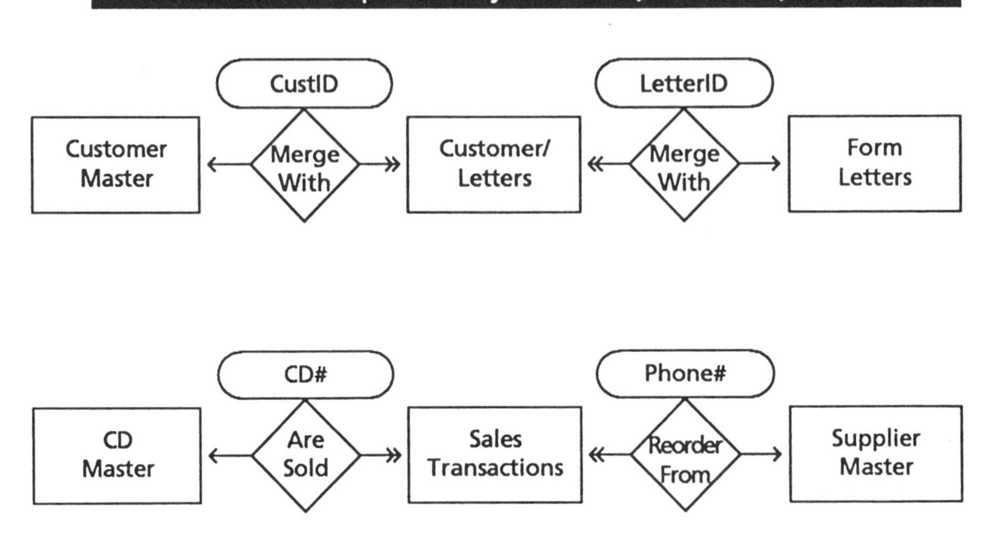

The New System File Design

Now that we have identified some of the file requirements for the new system, we can develop file structures that include specific field attributes (Figure 8-14). Nevertheless, this does not signal that the file design is complete. The report, query, and process design activities described in the next two chapters will almost certainly require some modifications to the new DFD and ERD models. This explains why these models were assigned version numbers.

The CD and Customer master files and the Sales transaction file will be implemented with relational database product. Once again, these structures are assigned a version number because we expect that changes will be introduced as we continue through the design process. The form letter file is a word processor file and as such is better described by a series of style sheets or templates. The Supplier file requires no formal computer structure, but the analyst should expect the user to develop a standard form for the entries in this file.

The New System Menu Tree

In keeping with our desires to inform and involve the user in the design process, a very simple menu tree (as shown in Figure 8-15) can lend some perspective to the series of choices that the user will have when the system is operational. Notice that in version 1.0 of this model, the Sales Trend subsystem is left blank. Details of this portion of the system await future JAD working sessions that occur during the report and query design activities.

Cornucopia New System File Structure (Version 1.0)

FIGURE 8–14

CUSTOMER Master File Structure

Field Name	Type	Size	Dec	Index	Comment
CNO	N	4		Y	Customer Number
LNAME	C	20		Y	Last Name
FNAME	C	20		N	First Name
STREET	C	20		N	
CITY	C	20		N	
STATE	C	2		N	
ZIPA	C	5		N	
ZIPB	C	4		N	
ACODE	C	3		N	
PREFIX	C	3		N	
PHNUM	C	4		N	
EDATE	D	8		N	Customer Since

CD Master File Structure

Field Name	Type	Size	Dec	Index	Comment
CDNO	C	12		Y	CD Number
TITLE	C	30		Y	
COMPOSER	C	30		Y	
ARTIST	C	30		Y	
LABEL	C	20		Y	
DESC	C	20		Y	
LPRICE	N	6	2	N	List Price
SUPPLIER	C	20		Y	

SALES Transaction File Structure

Field Name	Type	Size	Dec	Index	Comment
SDATE	D	8		Y	Sale Date
CDNO	C	12		Y	
SPRICE	N	6	2	Y	Sale Price
REORD	L	1		N	Reordered (Y/N)

Form Design

With the completion of the first versions of the new system models, the analyst and the user hold another JAD working session to develop the forms that will service the new system files. In preparation for this session, the analyst should create the database file structures and enter a small amount of sample data. This will allow the analyst and the user to actually create form prototypes using the screen-painting features common to most relational database packages. The following forms are designed to support the three major processing files in the system (Customer, CD, and Sales).

Figure 8-16 illustrates screen forms activated by the Customer Maintenance option on the Main menu and the Edit option on the Customer Maintenance menu. After the Customer file structure was created, these forms were created with the Create Form option in dBASE IV. Most relational database management products have similar screen-building tools.

Cornucopia New System Menu Tree (Version 1.0)

FIGURE 8–15

CORNUCOPIA INFORMATION SYSTEM
MAIN MENU
1. Sales Subsystem
2. Reorder Subsystem
3. Sales Trend Subsystem ...
4. Customer Maintenance ...
5. CD Maintenence ...
6. Correspondence ...
7. Exit
SALES TREND REPORTS MENU
1.
2.
3.
4. Return to Main Menu
CD MAINTENANCE MENU
1. View (Screen)
2. Print (Paper)
3. Edit (Add, Change, Delete)
4. Return to Main Menu
CUSTOMER MAINTENANCE
MENU
1. View (Screen)
2. Print (Paper)
3. Edit (Add, Change, Delete)
4. Return to Main Menu
CORRESPONDENCE MENU
1. Written Composition
2. Graphic Composition
3. Mail Merge
4. Print Customer Labels
5. Return to Main Menu

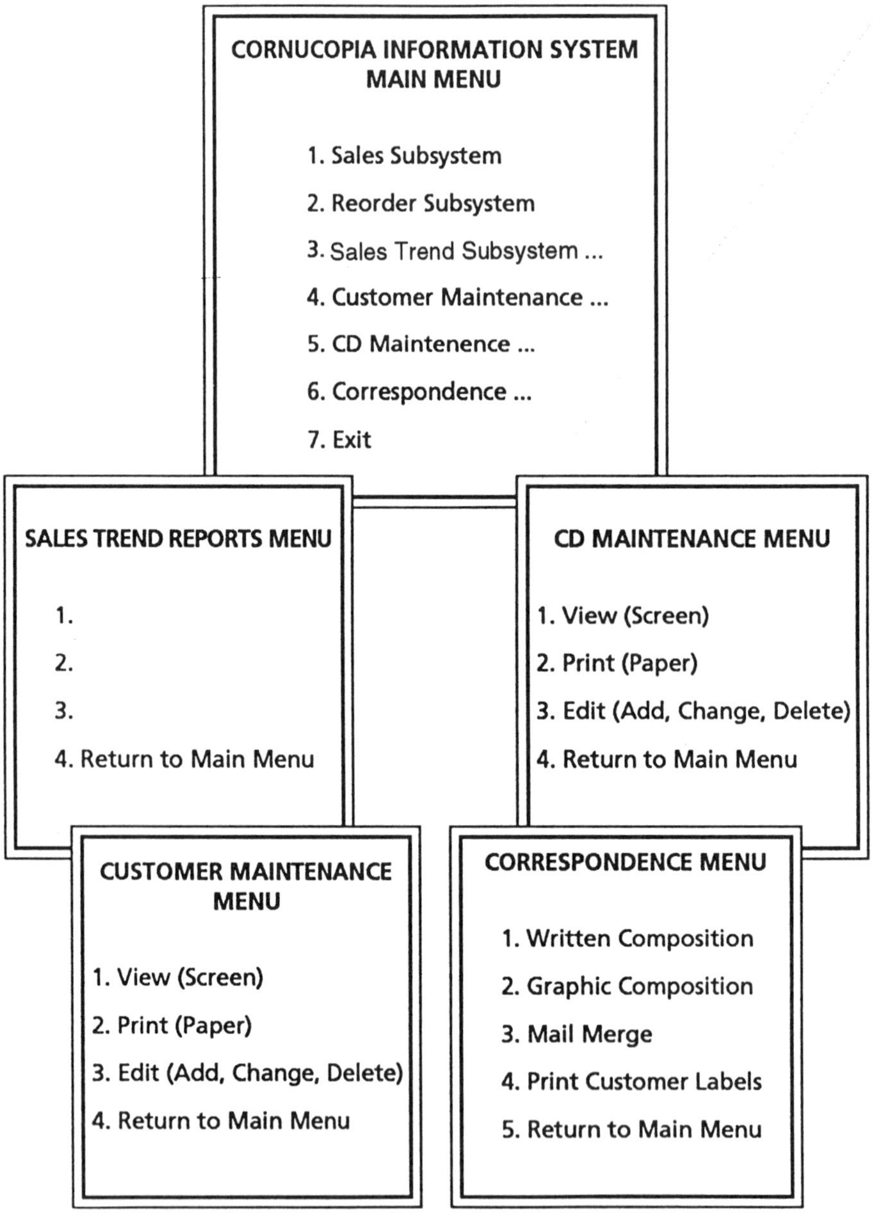

FIGURE 8–16 Cornucopia Customer Maintenance Screen Forms

```
                              CORNUCOPIA
DATE: 09/01/94          CUSTOMER MAINTENANCE SCREEN

   ADD        CHANGE              DELETE             EXIT

               CUSTOMER NUMBER: ....            SINCE: ../../..
   FIRST NAME: ..........................   LAST NAME: ....................
   STREET: ...........................   CITY: ....................
   STATE: ..  ZIP: .....-....  PHONE: (...)  ...-....
```

```
                              CORNUCOPIA
DATE: 09/01/94          CUSTOMER MAINTENANCE SCREEN

   ADD        CHANGE              DELETE             EXIT

               CUSTOMER NUMBER: 1122            SINCE: 12/20/93
   FIRST NAME: JANE                 LAST NAME: SMITH
   STREET: 123 MAPLE                CITY: EUREKA
   STATE: CA  ZIP: 95501-0000  PHONE: (707)  122-3444

               ACTION: U(pdate), C(hange), A(bort) U
               Add another customer? (Y/N) Y
```

```
                              CORNUCOPIA
DATE: 09/01/94          CUSTOMER MAINTENANCE SCREEN

   ADD        CHANGE              DELETE   Browse    EXIT
                                           Select

               CUSTOMER NUMBER: 1122            SINCE: 12/20/93
   FIRST NAME: JANE                 LAST NAME: SMITH
   STREET: 123 MAPLE                CITY: EUREKA
   STATE: CA  ZIP: 95501-0000  PHONE: (707)  122-3444

               ACTION: D(elete), B(rowse), A(bort) B
```

Figure 8-17 illustrates the forms required to maintain the CD file. Of course, both the Customer and CD files are master files. As a future upgrade, the CD file maintenance could be greatly enhanced by downloading new product updates through a telecommunications link. An even less complicated updating procedure might be a simple diskette driven system, in which the suppliers would provide the retailer with a new disk each month.

Cornucopia CD Maintenance Screen Forms

FIGURE 8–17

```
DATE: 09/01/94              CORNUCOPIA
                      CD MAINTENANCE SCREEN

   ADD        CHANGE        DELETE        EXIT

            Enter the CD number:

   CD NUMBER: ............       TITLE: ..................................
   COMPOSER:  ...........................
   ARTIST:  ............................  LABEL: ..............................
   DESCRIPTION: .....................  SUPPLIER: ...............................
   LIST PRICE: ......
```

```
DATE: 09/01/94              CORNUCOPIA
                      CD MAINTENANCE SCREEN

   ADD        CHANGE        DELETE        EXIT

            Enter the CD number: 234567890

   CD NUMBER: 234567890          TITLE: SAMPLE CD
   COMPOSER:  COMPOSER 1
   ARTIST:  ARTIST 1             LABEL: LABEL 1
   DESCRIPTION: TEST CASE        SUPPLIER:
   LIST PRICE: 14.95

         ACTION: U(pdate), C(hange screen), A(bort)  U
                  Add another CD? (Y/N)  Y
```

```
DATE: 09/01/94              CORNUCOPIA
                      CD MAINTENANCE SCREEN

   ADD        CHANGE  | Browse |  DELETE        EXIT
                      | Select |

            Enter the CD number: 234567890
            Proceed with your changes.

   CD NUMBER: 234567890          TITLE: SAMPLE CD
   COMPOSER:  COMPOSER 1
   ARTIST:  ARTIST 1             LABEL: LABEL 1
   DESCRIPTION: TEST CASE        SUPPLIER:
   LIST PRICE: 14.95

         ACTION: U(pdate), C(hange screen), A(bort)  C
```

Figure 8-18 presents the maintenance screens for the Sales transaction file. This involves a moderately complex series of dialogs between the user and the system. The careful reader here will notice that in addition to the basic data collection function, editing and pricing override features have been added to the design. To be fair, it should note that some dBASE programming code is used here to enhance the screen builder. Also, the Daily Summary and Update options are not illustrated in this figure. They are part of the report design activity covered in the next chapter.

FIGURE 8–18 Cornucopia Sales Transaction Screen Forms

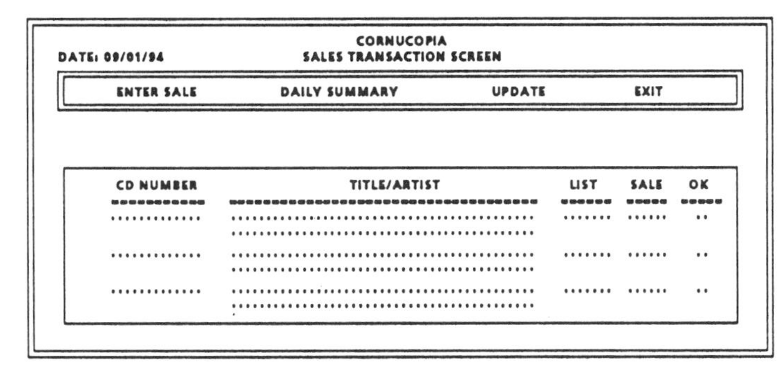

CORNUCOPIA
DATE: 09/01/94 SALES TRANSACTION SCREEN

ENTER SALE | DAILY SUMMARY | UPDATE | EXIT

CD NUMBER	TITLE/ARTIST	LIST	SALE	OK
....	....	....	......	..
....	....	....	......	..
....	....	....	......	..

CORNUCOPIA
DATE: 09/01/94 SALES TRANSACTION SCREEN

ENTER SALE | DAILY SUMMARY | UPDATE | EXIT

OK to process this transaction?

CD NUMBER	TITLE/ARTIST	LIST	SALE	OK
7863554022	BEETHOVEN & BRAHMS CONCERTOS HEIFETZ	15.99	15.99	OK
8940801582	SYMPHONY NO. 5 IN D MAJOR PREVIN	17.98	15.98	..
....	....	....	......	..

ACTION: P(rocess), A(bort) P

CORNUCOPIA
DATE: 09/01/94 SALES TRANSACTION SCREEN

ENTER SALE | DAILY SUMMARY | UPDATE | EXIT

CD NUMBER	TITLE/ARTIST	LIST	SALE	OK
7863554022	BEETHOVEN & BRAHMS CONCERTOS HEIFETZ	15.99	15.99	OK
8940801582	SYMPHONY NO. 5 IN D MAJOR PREVIN	17.98	15.98	OK
1234567891	Transaction aborted......................	....	......	..

The Project Dictionary

By this point in the project, there is a long list of items that are important to the overall project., including files, charts, forms, and diagrams. In the absence of a CASE tool or a project management software package, the analyst must develop a systematic catalog system to keep track of these project documents. Figure 8-19 presents the current entries in the database file used to maintain our dictionary. The file layout for such a database is presented in Chapter 6. Notice the standard naming convention devised to make the database easier to use (BUD stands for budget, PST stands for project status, etc).

Cornucopia Project Dictionary

FIGURE 8–19

DocID	DocName	SoftRef	Desc	Location
FRP01	FEASO1.DOC	WP	Feas. Rpt.	c:\wpdocs\corncpia
CNT01	CONT01.DOC	WP	Contract	c:\wpdocs\corncpia
CTX01	EXTX01.BMP	PB	Exst Context	c:\pbfigs\corncpia
DFD01	EDFD01.BMP	PB	Exst DFD-L1	c:\pbfigs\corncpia
USD01	EUSD01.BMP	PB	Exst USD	c:\pbfigs\corncpia
ERD01	EERD01.BMP	PB	Exst ERD	c:\pbfigs\corncpia
BUD03	CBUD03.WKS	QP	Budget Wk3	c:\qprwks\corncpia
PST03	SPST03.WKS	QP	PrjStat Wk3	c:\qprwks\corncpia
TSK01	CTSK01.DOC	WP	Task List	c:\wpdocs\corncpia
PWK01	CPWK01.DOC	WP	PERT Wksht	c:\wpdocs\corncpia
PRT01	CPRT01.BMP	PB	PERT Chart	c:\pbfigs\corncpia
PRE01	CPRE01.DOC	WP	Prj Prelim	c:\wpdocs\corncpia
USD11	CUSD11.BMP	PB	New USD v1	c:\pbfigs\corncpia
DFD11	CDFD11.BMP	PB	New DFD v1	c:\pbfigs\corncpia
ERD11	CERD11.BMP	PB	New ERD v1	c:\pbfigs\corncpia
MEN11	CMEN11.BMP	PB	New MTree v1	c:\pbfigs\corncpia
DAT01	CUSTMF.DBF	DB	Cust Master	c:\dbase4\corncpia
DAT02	CDMF.DBF	DB	CD Master	c:\dbase4\corncpia
DAT03	SALETF.DBF	DB	Sales Trans	c:\dbase4\corncpia
SCR01	CUSTFM.SCR	DB	Cust Form	c:\dbase4\corncpia
SCR02	CDFM.SCR	DB	CD Form	c:\dbase4\corncpia
SCR03	SALEFM.SCR	DB	Sale Form	c:\dbase4\corncpia
PDT01	CPDT01.DBF	DB	Prj Dict	c:\dbase4\corncpia
BUD05	CBUD05.WKS	QP	Budget Wk5	c:\qprwks\corncpia
PST05	CPST05.WKS	QP	PrjStat Wk5	c:\qprwks\corncpia

Time and Money

The analyst reports 8 hours against this project during the week, all of which were spent on design. Figure 8-20 shows the project status as of Week 5. Notice that the actual hours reported to the design activities total more than 50 percent of the estimated total, but the percentage complete is only 30 percent. This illustrates the difficulty of estimating systems work. Overall, the project is still within budget for labor because the actual hours required to do the analysis are comfortably less than the estimates.

FIGURE 8–20 Cornucopia Project Status—Week 5

Date: | **Cornucopia Project Status** | **A/O Week: 5**

Activity	% Comp.	Status	1	2	3	4	5	6	7	8	9	0	1	2	3	4	5	6	Total
Analysis-Est	100%		4	5	5	4													18
Actuals	95%	ok	2	4	2	3													11
Design-Est	30%					4	4	5	5	4									22
Actuals	30%	ok			1	3	8												12
Develop-Est	0%									4	4	5	5	4					22
Actuals	0%	ok																	0
Impl.-Est	0%													4	5	5	5	5	24
Actuals	0%	ok																	0
Total-Est			4	5	5	8	4	5	5	8	4	5	5	8	5	5	5	5	86
Actuals			2	4	3	6	8	0	0	0	0	0	0	0	0	0	0	0	23
Contract	100%	ok	C																
Prelim. Pres.	100%	ok			C														
Design Rev.	0%	ok								S									
Proto. Rev.	0%	ok											S						
Train & Del.	0%	ok															S		
Final Rpt.	0%	ok																S	
			1	2	3	4	5	6	7	8	9	0	1	2	3	4	5	6	

Chapter 9
Report and Query Design

The JAD sessions devoted to output design begin with a series of standard outputs that are based on the data stores identified in the new system DFD (see Figure 8-12). The project contract and preliminary presentation are also referred in order to determine other basic output requirements. Not surprisingly, most of the serious design work focuses on the sales trend reports. Figure 9-9 presents the list of outputs for the Cornucopia project. Notice that the list is organized into content groupings, within which are the different presentation and format descriptions.

Cornucopia Output List **FIGURE 9–9**

Output Name	Report/Query	Schedule/On-Demand	Hardcopy/Softcopy
Master Files			
1. Customers	Report	On-demand	Hard/soft
2. CDs	Report	On-demand	Hard/soft
Transaction Files			
3. Sales Transactions	Report	On-demand	Hard/soft
4. Sales Summary	Report	On-demand	Soft
5. Sales Inquiry	Inquiry	On-demand	Soft
Sales Trends			
6. Last 30 Days	Report	On-demand	Hard
7. Last 12 Months	Report	On-demand	Hard
8. Year to Year	Report	On-demand	Hard
9. Units/Price	Report	On-demand	Hard
10. Amount/Price	Report	On-demand	Hard
11. Monthly Sales	Report	Scheduled	Hard
12. Amount/Unit Query	Query	On-demand	Soft
Reorder			
13. Reorder Report	Report	On-demand	Hard
Correspondence			
14. Letters	n/a	On-demand	Hard
15. Newsletter	n/a	On-demand	Hard
16. Labels	n/a	On-demand	Hard

New System Output Requirements

The list of outputs includes reports for the computer-based Customer and CD master file updates and the Sales transaction file activity (items 1-3). To save paper and storage space, the user agrees that these reports should be produced only on-demand and should be available as either hardcopy or softcopy.

The Sales Summary (item 4) is a softcopy, on-demand report available through the Sales subsystem menu choice Daily Summary. The Sales Inquiry (item 5) gives the user access to the Sales transaction file through SQL commands. Both these outputs were designed during the JAD session, when the user considered the type of unplanned information needs that might occur during the business day.

The sales trends reports (items 6-11), first described in general terms in the project contract, appear on the list in much more detail. As described in the following material, two new files and some changes to the sales transaction file structure are required to support these reports. The Unit Query (item 12) allows the user to generate historical sales transaction information through a user-friendly query form. The CD Reorder Report (item 13) provides a list of daily sales, with space for the owner to record order and receipt information.

The correspondence outputs (items 14-16) are not illustrated.

The Revised New System Models

The output design activities developed the specific details for the content, presentation and format for the sales trends reports. This presents a practical question: How can the analyst be so far along in the new system design without having developed such detail long ago? The answer lies in the very nature of problem solving and systems work. The ***enhanced*** SDLC model is circular because information systems continue to evolve even after the initial project definition work completes. Further, the individual activity phases of the model tend to blend together such that it is sometimes difficult to pinpoint the current phase. Finally, the analyst will often revisit specific activities within an earlier phase because of work accomplished in a later phase.

Working like this resembles the way an artist works by painting the background and blocking the composition before adding the detail. In systems work this approach is inevitable. Because it is impossible to visualize the entire project at once, so we split the problem into more manageable parts and then solve them one at a time, with the understanding that we may need to go back to fill in some of the details from time to time.

The revised menu tree appears in Figure 9-10. It reflects the new detail concerning the sales trends reports.

The revised data flow diagram appears in Figure 9-11. It also reflects new detail about the data stores and data flows required to support the sales trends reports. The user's desire to see sales amounts and units sold summarized by pricing ranges requires some new fields in the sales transaction file. Also, the historical summaries (items 9-11) require new files (daily and monthly history) to retain sales data in terms of these new categories.

Cornucopia New System Menu Tree (Version 1.1)

FIGURE 9–10

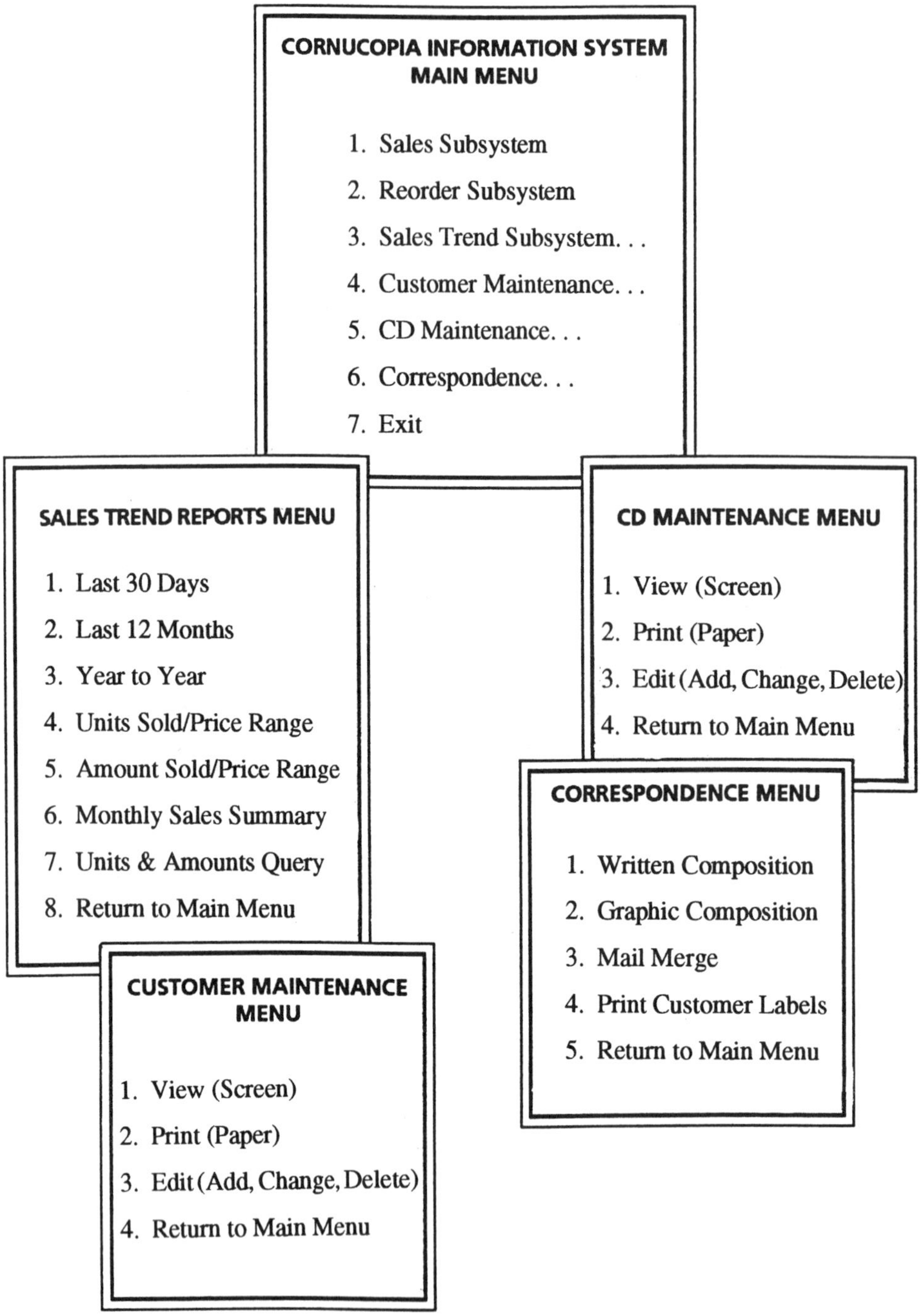

CORNUCOPIA INFORMATION SYSTEM
MAIN MENU

1. Sales Subsystem
2. Reorder Subsystem
3. Sales Trend Subsystem. . .
4. Customer Maintenance. . .
5. CD Maintenance. . .
6. Correspondence. . .
7. Exit

SALES TREND REPORTS MENU

1. Last 30 Days
2. Last 12 Months
3. Year to Year
4. Units Sold/Price Range
5. Amount Sold/Price Range
6. Monthly Sales Summary
7. Units & Amounts Query
8. Return to Main Menu

CD MAINTENANCE MENU

1. View (Screen)
2. Print (Paper)
3. Edit (Add, Change, Delete)
4. Return to Main Menu

CORRESPONDENCE MENU

1. Written Composition
2. Graphic Composition
3. Mail Merge
4. Print Customer Labels
5. Return to Main Menu

CUSTOMER MAINTENANCE MENU

1. View (Screen)
2. Print (Paper)
3. Edit (Add, Change, Delete)
4. Return to Main Menu

FIGURE 9–11 **Cornucopia New System First-Level DFD (Version 1.1)**

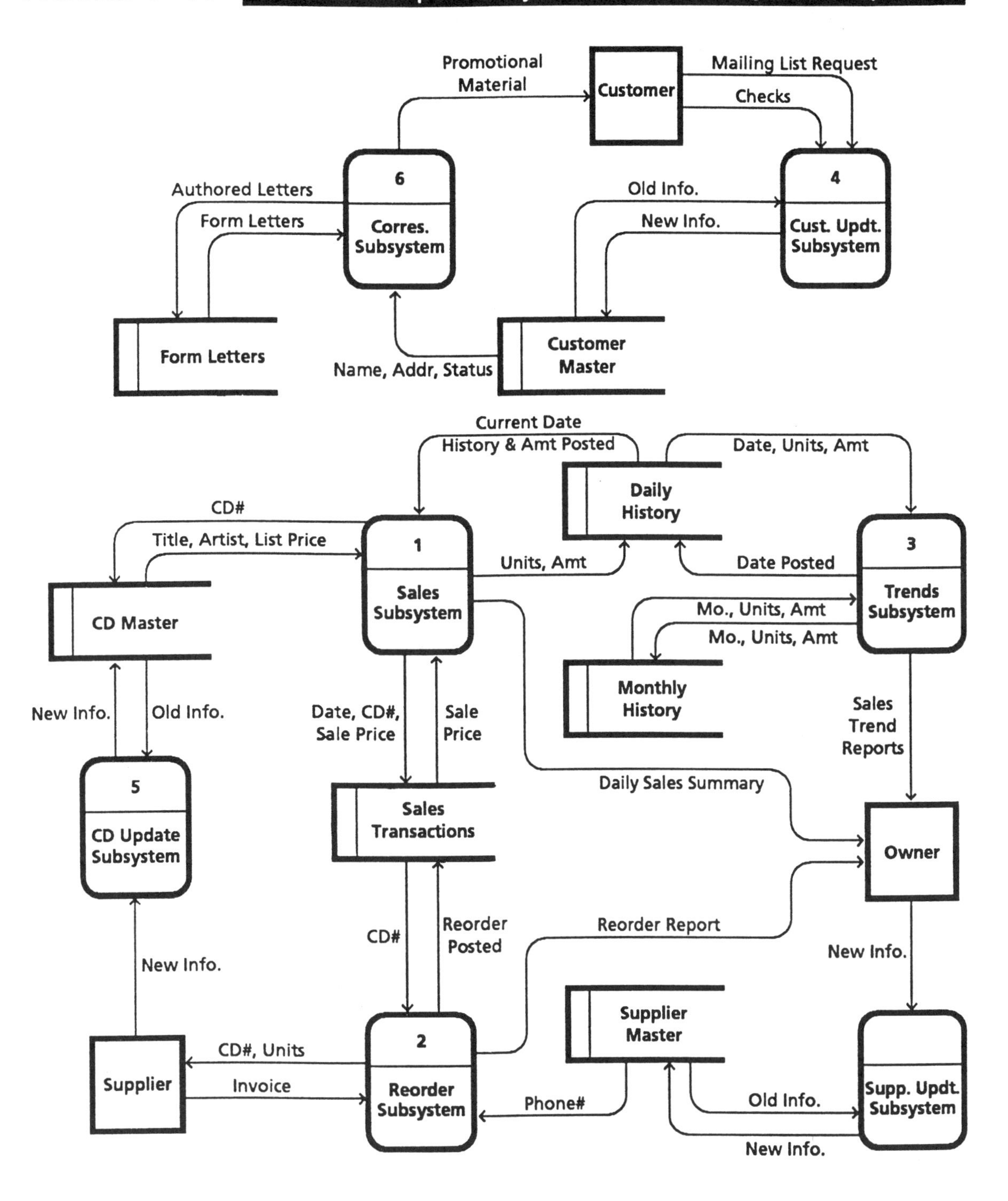

Report Design

Sample report designs appear in the following figures. Notice that all but two reports (items 3 and 11) are actual screen images produced from the prototypes.

FIGURE 9–12 **Cornucopia Master File Reports**

Item 1

CORNUCOPIA
CUSTOMER REPORT

DATE: PAGE: 1

Name/Address		Since	Phone	ID
PHIL	JOHNSON	08/01/92	(707) 433-1234	1111
4567 HOWARD ST.				
EUREKA, CA	95501-0000			
JUNE	ALTON	08/03/92	(707) 883-4567	1112
234 FRANCIS CT.				
ARCATA, CA	95512-0000			
MASON	KIRTLAND	07/15/92	(707) 768-5678	1113
P.O. BOX 654				
LOLETA, CA	95506-0654			
GERGIO	MELIZE	08/24/92	(707) 725-1234	1114
56 OAK ST.				
FORTUNA, CA	95540-0000			
JULIE	SQUIRT	08/24/92	(707) 725-8989	1115
4321 GOLDEN HEIGHTS				
FORTUNA, CA	95540-0000			
HARRY	WALKER	09/12/92	(707) 443-6767	1116
1212 "C" ST.				
EUREKA, CA	95502-0000			
DAVID	TEMPLETON	09/15/92	(707) 667-1212	1117
4567 BEACON WAY				
FIELDS LANDING, CA	95507-0000			
GEORGE	SMITH	08/28/92	(707) 443-6789	1118
6789 "G" ST.				
EUREKA, CA	95501-0000			
FRANCIS	LOWE	09/01/92	(707) 883-1234	1119
4567 WALTER BLVD.				
ARCATA, CA	95508-0000			

Item 2

CORNUCOPIA
CD REPORT

DATE: PAGE: 1

Composer/Title/Artist/Label	CD Number/Description	List Price/Supplier
BEETHOVEN,BRAHMS	7863554022	15.99
BEETHOVEN & BRAHMS CONCERTOS	VIOLIN	
HEIFETZ		
RCA		
WILLIAMS	8940801582	17.98
SYMPHONY NO. 5 IN D MAJOR	ORCHESTRAL	
PREVIN		
TELARC		

Cornucopia Sales Transaction File Reports

FIGURE 9–13

Item 3

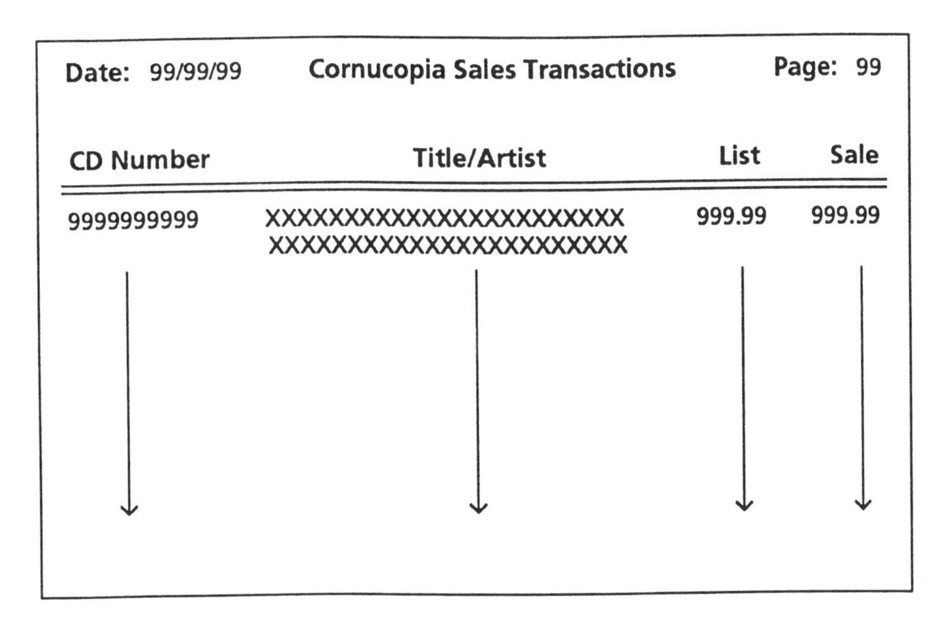

Date: 99/99/99 Cornucopia Sales Transactions Page: 99

CD Number	Title/Artist	List	Sale
9999999999	XXXXXXXXXXXXXXXXXXXXXXX XXXXXXXXXXXXXXXXXXXXXXX	999.99	999.99

Item 4

CORNUCOPIA
SALES TRANSACTION SCREEN

DATE: 12/20/93

ENTER SALE   DAILY SUMMARY   UPDATE   EXIT

DAILY SALES SUMMARY   11:07:59

UNITS: 2

AMOUNT: 31.97

Press any key to continue...

FIGURE 9–14 Cornucopia Sales Trend Reports

Item 6

DATE: 12/20/93 — CORNUCOPIA SALES TRENDS — LAST 30 DAYS — PAGE: 1

Last 30 Days	Units Sold	Amount Sold
12/13/93	33	585
12/14/93	38	684
12/15/93	35	639
12/16/93	37	621
[illegible]	39	666
		666

Item 7

DATE: 12/20/93 — CORNUCOPIA SALES TRENDS — LAST 12 MONTHS — PAGE: 1

Last 30 Days	Units Sold	Amount Sold
12 / 1992	1,174	19,278
1 / 1993	248	3,789
2 / 1993	197	2,772
3 / 1993	300	4,860
4 / 1993	281	4,473
[illegible]	320	5,733
	[illegible]	2,457
		[illegible]

Item 8

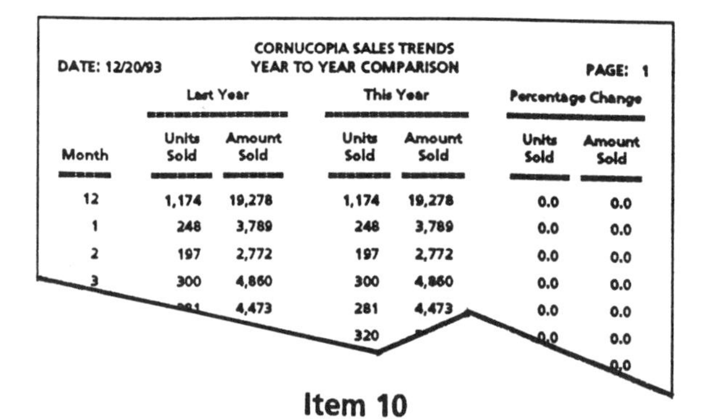

DATE: 12/20/93 — CORNUCOPIA SALES TRENDS — YEAR TO YEAR COMPARISON — PAGE: 1

Month	Last Year Units Sold	Last Year Amount Sold	This Year Units Sold	This Year Amount Sold	Percentage Change Units Sold	Percentage Change Amount Sold
12	1,174	19,278	1,174	19,278	0.0	0.0
1	248	3,789	248	3,789	0.0	0.0
2	197	2,772	197	2,772	0.0	0.0
3	300	4,860	300	4,860	0.0	0.0
	[illegible]	4,473	281	4,473	0.0	0.0
			320	[illegible]	0.0	0.0
						0.0

Item 9

DATE: 12/20/93 — CORNUCOPIA SALES TRENDS — UNITS SOLD PER PRICE RANGE — PAGE: 1

Last 12 Months	Percentage of Units Sold: Less than $11	$11 to $20	More than $20
12 / 1992	44.5	28.4	27.0
1 / 1993	50.4	29.4	20.2
2 / 1993	60.9	21.8	17.3
3 / 1993	43.3	33.3	23.3
4 / 1993	45.6	32.0	22.4
	31.3	38.4	30.3
	49.4	[illegible]	22.2
			[illegible]

Item 10

DATE: 12/20/93 — CORNUCOPIA SALES TRENDS — UNITS SOLD PER PRICE RANGE — PAGE: 1

Last 12 Months	Percentage of Sale Amounts: Less than $11	$11 to $20	More than $20
12 / 1992	24.4	31.2	44.4
1 / 1993	29.7	34.7	35.6
2 / 1993	39.0	27.9	33.1
3 / 1993	24.1	37.0	38.9
[illegible]	25.8	36.2	38.0
	15.7		45.7
			[illegible]

Item 11

Date: 99/99/99 — Cornucopia Monthly Sales — Month: XXXXXXXXXX — Page: 99

Price Range	Units	%	Amounts	%
< $11	9,999	99.9	99,999.99	99.9
$11 to $20	9,999	99.9	99,999.99	99.9
> $20	9,999	99.9	99,999.99	99.9
Totals	99,999		999,999.99	

FIGURE 9–15 Cornucopia CD Reorder Report

Item 13

DATE: MM/DD/YY	CORNUCOPIA CD REORDER REPORT			PAGE: 99
Supplier: XXXXXXXXXXXXXXXXXXX				
CD Number/Title	Qty Sold	Ordered	Received	Comment
99999999999 XXXXXXXXXXXXXXXXXXXXXX	99999	....	....	....
99999999999 XXXXXXXXXXXXXXXXXXXXXX	99999	....	....	....
99999999999 XXXXXXXXXXXXXXXXXXXXXX	99999	....	....	....
99999999999 XXXXXXXXXXXXXXXXXXXXXX	99999	....	....	....

Query Design

The Sales Inquiry (item 5 in Figure 9-16) illustrates two sample SQL statements. The user understands that this type of output is available only through the SQL interface, which requires special syntax training and knowledge of the database schema. Although it is unlikely that this access method will be used as a regular part of the information system, the user is interested to know that it is an option.

The Units & Amounts Query (item 12 in Figure 9-16) is much more important to the user at this time. This type of on-demand, softcopy access to the historical sales amounts and units sold data allows the user to focus on one particular area of sales/unit information, unlike the larger summaries provided by the sales trends reports. The fact that this form is already designed and easily used is a comfort to the user.

FIGURE 9–16 Cornucopia Sales Inquiry and Query

Item 5

Cornucopia Sales Inquiry
Same Entries
.LIST LAST 10
(shows last 10 sales)
.LIST FOR cdno = 1234567890
(shows sales for specific CD)
Type RETURN to exit.

Item 12

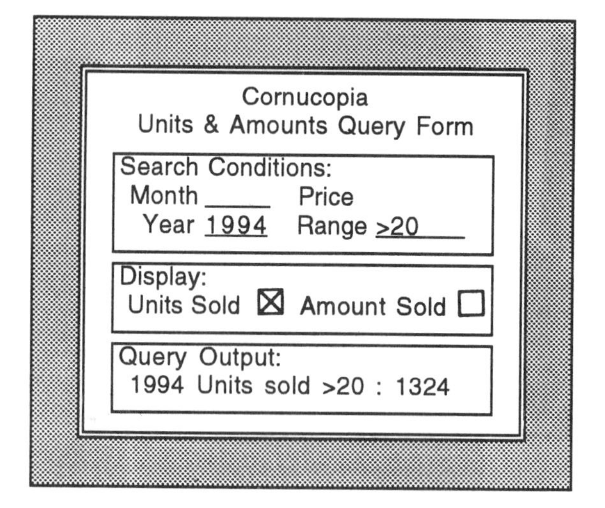

I/O System Resource Requirements

The input/output resource requirements cannot be finalized until the process design activities are well underway. Nevertheless, identifying the general nature of these requirements as the design phase progresses is very important. The input/output design decisions are often made with specific hardware and software in mind, which can, in-turn, affect the process design itself. Figure 9-17 proposes several generic hardware and software items that are required to support the input and output design to date.

Cornucopia I/O System Resource Requirements Notes **FIGURE 9–17**

Hardware

Processing Platform:

The intended use of a database processing package for transaction processing and master file maintenance requires a minimum platform processor speed of 33MHz, with at least 200Mb of had disk storage.

Peripherals:

The sales transaction processing subsystem should be supported with a bar code scanner to minimize input time and error rate.

The modest amount of hardcopy output (reports and correspondence) suggests a single laser quality, black-and-white printer.

Software

Relational database and word processing software is required. The need for spreadsheet and graphics software is still in question.

Time and Money

A total of 5 hours is reported against the design activities for this project during the past week. The analyst estimates that the design is now 50 percent complete.

Figures 9-18 and 9-19 illustrate the project budget and status as of Week 6. Notice that the hardware and software purchases are initiated during this period, to allow sufficient time for delivery, installation, and testing before the project implementation activities commence in Week 12. These purchases generally follow the resource requirement notes shown in Figure 9-17. A detailed itemization appears in Chapter 11.

FIGURE 9–18 **Cornucopia Project Budget—Week 6**

Date: Cornucopia Budget A/O Wk: 6

	Week1	Week2	Week3	Week4	Week5	Week6	Week7	Week8	Week9	Week10	Week11	Week12	Week13	Week14	Week15	Week16	Total
Estimates																	
Hardware						2500	500	500	500								4000
Software						1000	250	250									1500
Labor	200	250	250	400	200	250	250	400	200	250	250	400	250	250	250	250	4300
Total	200	250	250	400	200	3750	1000	1150	700	250	250	400	250	250	250	250	9800
Actuals																	
Hardware						1995											1995
Software						495											495
Labor	100	200	150	300	400	250											1400
Total	100	200	150	300	400	2740	0	0	0	0	0	0	0	0	0	0	3890
Weekly +/–																	
Hardware	0	0	0	0	0	505	0	0	0	0	0	0	0	0	0	0	505
Software	0	0	0	0	0	505	0	0	0	0	0	0	0	0	0	0	505
Labor	100	50	100	100	-200	0	0	0	0	0	0	0	0	0	0	0	150
Total	100	50	100	100	-200	1010	0	0	0	0	0	0	0	0	0	0	1160
Cumm. +/–																	
Hardware	0	0	0	0	0	505	0	0	0	0	0	0	0	0	0	0	
Software	0	0	0	0	0	505	0	0	0	0	0	0	0	0	0	0	
Labor	100	150	250	350	150	150	0	0	0	0	0	0	0	0	0	0	
Total	100	150	250	350	150	1160	0	0	0	0	0	0	0	0	0	0	

Cornucopia Project Status — Week 6

FIGURE 9-19

Date: Cornucopia Project Status A/O Week: 6

Activity	% Comp.	Status	1	2	3	4	5	6	7	8	9	0	1	2	3	4	5	6	Total
Analysis-Est	100%		4	5	5	4													18
Actuals	100%	ok	2	4	2	3													11
Design -Est	60%					4	4	5	5	4									22
Actuals	50%	ok			1	3	8	5											17
Develop-Est	0%									4	4	5	5	4					22
Actuals	0%	ok																	0
Impl. -Est	0%													4	5	5	5	5	24
Actuals	0%	ok																	0
Total -Est			4	5	5	8	4	5	5	8	4	5	5	8	5	5	5	5	86
Actuals			2	4	3	6	8	5	0	0	0	0	0	0	0	0	0	0	28
Contract	100%	ok	C																
Prelim. Pres.	100%	ok			C														
Design Rev.	0%	ok								S									
Proto. Rev.	0%	ok											S						
Train & Del.	0%	ok															S		
Final Rpt.	0%	ok																S	
			1	2	3	4	5	6	7	8	9	0	1	2	3	4	5	6	

Chapter 10
Process Design

As you will recall, the owner feels that a "point-of-sale/inventory/reorder" type of system is too complicated and expensive at this time. She wants to take this slowly. As a compromise, the system will use a bar code scanner to capture the universal product code (UPC) of the compact disks (CDs) that are sold. The existing cash register procedure remains unchanged. Data capture procedures for other products will be added at a later date.

The four areas of concern (customer recordkeeping, reordering system, customer communications, and sales trending) are all addressed in the design presented in the revised DFD, USD, and menu tree. The six processes in this system are listed in Figure 10-14.

Cornucopia Processing Requirements **FIGURE 10–14**

Process	Implementing Software
Customer update	Database
CD update	Database
Sales transactions	Database
Reordering	Database
Sales tending	Database Spreadsheet
Correspondence	Word processing Database

The Revised New System Data Model

The report and query design decisions concerning the sales trend outputs affect the new system data model. Specifically, the cumulative nature of the reports requires a summarization of the daily sales transactions. Because this was not addressed during the initial file-design activities, we must develop these designs now.

This exercise illustrates that system design does not always proceed in a sequential fashion. If the analyst has carefully documented the design at each stage of its development, the redesign is much easier. For example, the original new system ERD (from Figure 8-13) is simply revised to include the history files and to indicate that the Supplier master file will remain a manual operation. Figure 10-15 shows the revised ERD.

FIGURE 10–15 **Cornucopia New System ERD (Version 1.1)**

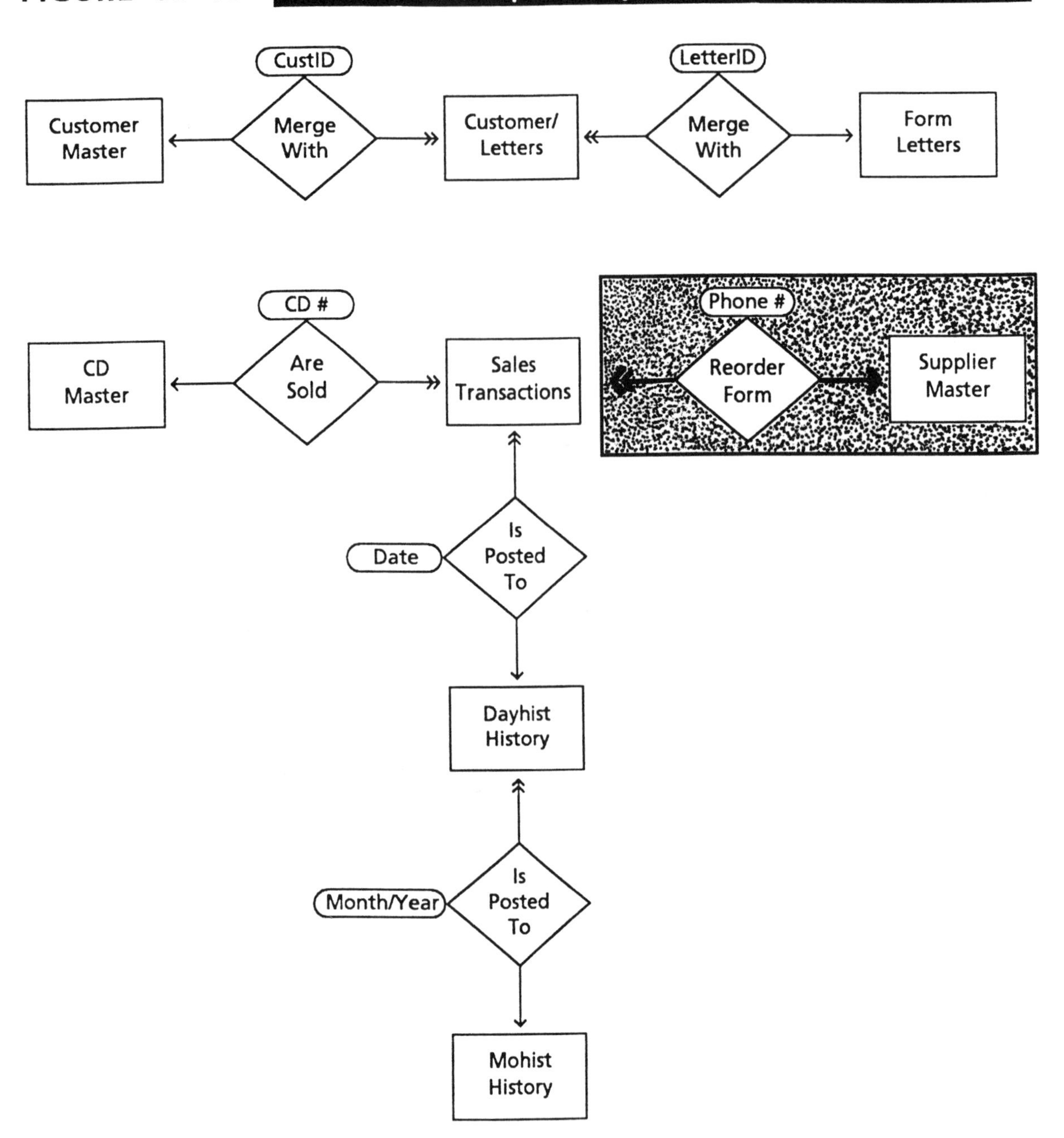

Along with the revised ERD, the file structures are also revised to add the two new files and to include some minor changes in the remaining master files and the Sales transaction file. The status field has been added to the master files so that the system can record the CDs removed from supplier files customers the owner wishes to classify as no longer active. The user suggested this change because she wants to be able to communicate with old customers and identify old CDs. The posted field was added to the Sales transaction file so that the update routine can set a filter to process only those sales that have not been posted.

FIGURE 10–16 Cornucopia New System File Structures (Version 1.1)

CUSTOMER Master File Structure

Field Name	Type	Size	Dec	Index	Comment
CNO	N	4		Y	Customer Number
LNAME	C	20		Y	Last Name
FNAME	C	20		N	First Name
STREET	C	20		N	
CITY	C	20		N	
STATE	C	2		N	
ZIPA	C	5		N	
ZIPB	C	4		N	
ACODE	C	3		N	
PREFIX	C	3		N	
PHNUM	C	4		N	
EDATE	D	8		N	Customer Service
STATUS	C	1		N	Active or Inactive

CD Master File Structure

Field Name	Type	Size	Dec	Index	Comment
CDNO	C	12		Y	CD Number
TITLE	C	30		Y	
COMPOSER	C	30		Y	
ARTIST	C	30		Y	
LABEL	C	20		Y	
DESC	C	20		Y	
LPRICE	N	6		N	List Price
SUPPLIER	C	20	2	Y	
STATUS	C	1		N	Active or Inactive

SALES Transaction File Structure

Field Name	Type	Size	Dec	Index	Comment
SDATE	D	8		Y	Sale Date
CDNO	C	12		Y	
SPRICE	N	6	2	Y	Sale Price
REORD	L	1		N	Reordered (Y/N)
SPOSTED	L	1		N	Posted to Day History

DAYHIST History File Structure

Field Name	Type	Size	Dec	Index	Comment
HDATE	D	8		Y	
DAYUNITS1	N	3		N	< $11
DAYUNITS2	N	3		N	$11 to $20
DAYUNITS3	N	3		N	> $20
DAYAMT1	N	4		N	
DAYAMT2	N	4		N	
DAYAMT3	N	4		N	
DPOSTED	C	1		N	Posted to Month History

MOHIST History File Structure

Field Name	Type	Size	Dec	Index	Comment
HMONTH	N	2		Y	
HYEAR	N	4		Y	< $11
MOUNITS1	N	4		N	$11 to $20
MOUNITS2	N	4		N	> $20
MOUNITS3	N	4		N	
MOAMT1	N	5		N	
MOAMT2	N	5		N	
MOAMT3	N	5		N	

The New System Flowchart

The system flowchart associates the real-world application programs, file types, and peripheral devices that are used to implement the design. It also shows the relationships between and among the data files and the processing subsystems. Both the DFD and the ERD are consulted to prepare such a flowchart. The analyst should be able to correlate the system flowchart items with these two models. In this way, the project design is reviewed for consistency and completeness before more detailed process design activities begin.

Cornucopia System Flowchart

FIGURE 10–17

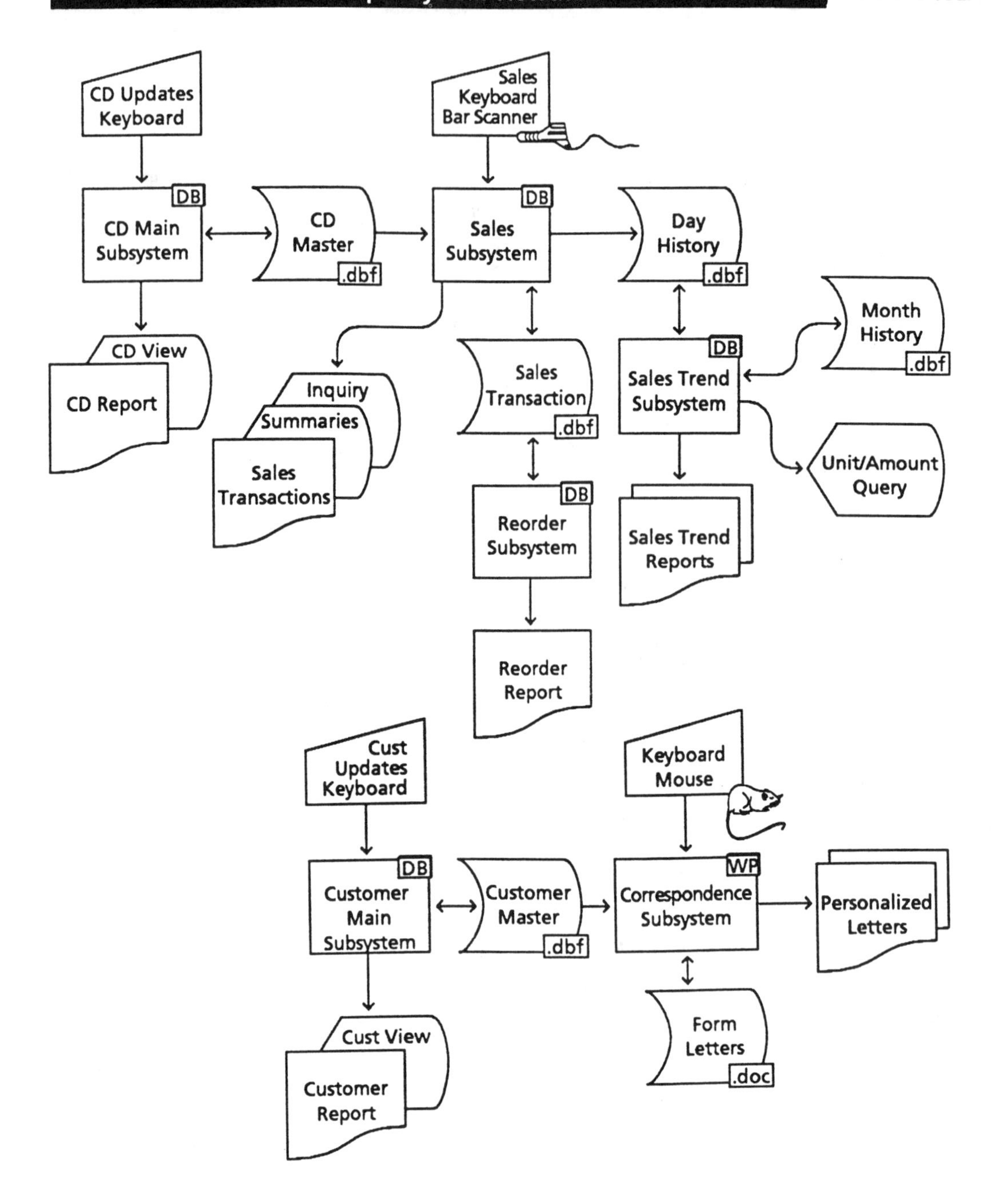

The Subsystem Structure Charts

Figures 10-18 and 10-19 present the subsystem structure charts for the master file maintenance and sales transaction processing logic. These charts show the modular composition of the subsystems, where each module performs a simple task. These modules are closely tied to the screen-form designs developed earlier. For example, the Customer master file maintenance screen design (Figure 8-16) calls for three menu options: Add, Change and Delete. Notice that these correspond to the CustAdd, CustChg, and CustDel modules in the structure chart. Similarly, the Sales transaction subsystem structure chart is patterned after the screen design (Figure 8-18).

FIGURE 10–18 **Cornucopia Master File Maintenance Subsystem Structure Chart**

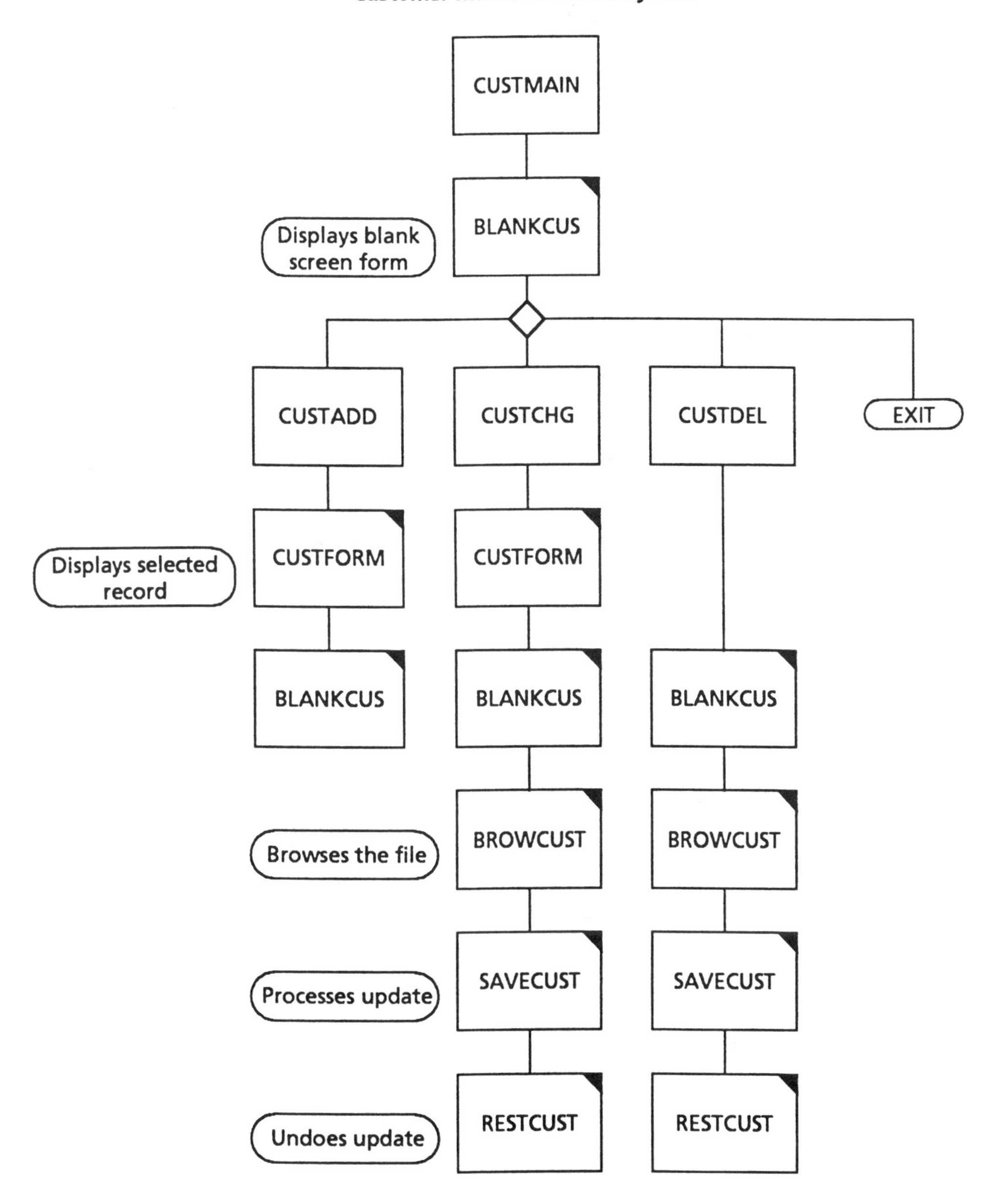

Cornucopia Sales Subsystem Structure Chart

FIGURE 10–19

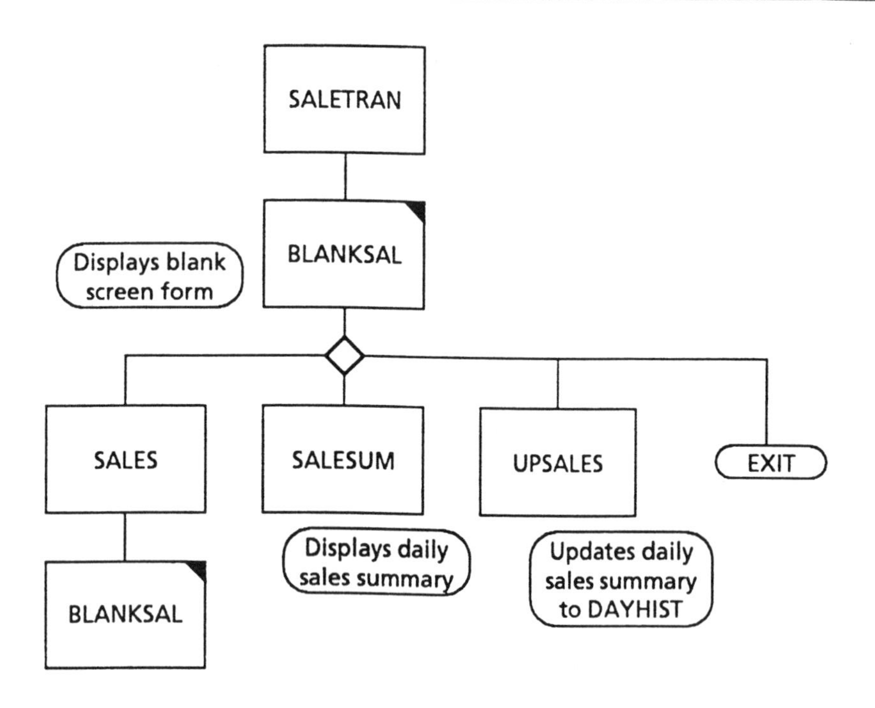

General Process Design

The Customer and CD master files are created and maintained using database software. These are batch operations in the sense that they are performed periodically, as required by new or changing data. They do not interrupt the normal operations of the enterprise. The Sales transaction file is updated whenever a CD is sold, which does add a small amount of work.

The assumption is that the user will have the OLTP sales subsystem (SALTRAN) "up" during business hours. Each sale will be recorded as it occurs (SALES). A summary of daily sales (SALESUM) can be called for through this subsystem. At the end of the day, the user can perform the update (UPSALES) to the Daily History file (DAYHIST).

The Sales Trends subsystem uses the DAYHIST file to generate the Last 30 Days report and to update the Monthly History file (MOHIST). Upon demand, the user can use the Sales Trends subsystem to create the other trend reports, which are based on the MOHIST file.

The Reordering system is very simple, producing a listing that will be used for phone orders to vendors and to record shipments. This subsystem is a candidate for future enhancement, perhaps with an electronic reordering and inventory system.

The Correspondence and Customer Maintenance (CUSTMAIN) subsystems stand apart from the Sales subsystem. This design should also be revisited later.

This is a multifile system that requires safeguards to ensure the referential integrity of the database. For example, every entry in the SALES.DBF must have a single corresponding entry in the CD.DBF. Further, the master files require safeguards as well.

Processing Resource Requirements

The processing design decisions confirm that the hardware specified in the I/O System Resource Requirements Notes (Figure 9-17) are appropriate. Based on the system flowchart, neither graphics nor spreadsheet software is needed.

One question persists however. The proposed keyboard-based update process for the CD master file promises to be very tedious and error prone. If you recall, the analyst's earlier research into vertical solutions produced a reasonably priced point-of-sale product called Phono-Scan. A part of this system included electronic updates for the CD master file. This suggests that an outside service can provide update information for the CD file. Based on the assumption that the user may very likely choose to upgrade to include this feature, the analyst recommends that a fax-modem be added to the hardware requirements as an optional feature.

Time and Money

The process design activities have consumed 7 hours of analyst time during the reporting period, bringing the estimated completion of this phase to 80 percent. Figure 10-20 illustrates the Project Status Report as of Week 7.

Cornucopia Project Status—Week 7 **FIGURE 10–20**

Date: Cornucopia Project Status A/O Week: 7

Activity		% Comp.	Status	1	2	3	4	5	6	7	8	9	0	1	2	3	4	5	6	Total
Analysis	-Est	100%		4	5	5	4													18
	Actuals	100%	ok	2	4	2	3													11
Design	-Est	80%					4	4	5	5	4									22
	Actuals	80%	ok			1	3	8	5	7										24
Develop	-Est	0%									4	4	5	5	4					22
	Actuals	0%	ok																	0
Impl.	-Est	0%													4	5	5	5	5	24
	Actuals	0%	ok																	0
Total	-Est			4	5	5	8	4	5	5	8	4	5	5	8	5	5	5	5	86
	Actuals			2	4	3	6	8	5	7	0	0	0	0	0	0	0	0	0	35
Contract		100%	ok	C																
Prelim. Pres.		100%	ok			C														
Design Rev.		0%	ok								S									
Proto. Rev.		0%	ok											S						
Train & Del.		0%	ok															S		
Final Rpt.		0%	ok																S	
				1	2	3	4	5	6	7	8	9	0	1	2	3	4	5	6	

Figure 10-21 illustrates the Project Budget as of Week 7. Notice that the software and hardware purchases continued during this period. As the process design progresses, the analyst is eager to order these products to allow sufficient time for shipping, installation, testing, and training.

FIGURE 10–21 Cornucopia Project Budget—Week 7

Date: Cornucopia Budget A/O Wk: 7

	Week1	Week2	Week3	Week4	Week5	Week6	Week7	Week8	Week9	Week10	Week11	Week12	Week13	Week14	Week15	Week16	Total
Estimates																	
Hardware						2500	500	500	500								4000
Software						1000	250	250									1500
Labor	200	250	250	400	200	250	250	400	200	250	250	400	250	250	250	250	4300
Total	200	250	250	400	200	3750	1000	1150	700	250	250	400	250	250	250	250	9800
Actuals																	
Hardware						1995	595										2590
Software						495	149										644
Labor	100	200	150	300	400	250	350										1750
Total	100	200	150	300	400	2740	1094	0	0	0	0	0	0	0	0	0	4984
Monthly +/–																	
Hardware	0	0	0	0	0	505	-95	0	0	0	0	0	0	0	0	0	410
Software	0	0	0	0	0	505	101	0	0	0	0	0	0	0	0	0	606
Labor	100	50	100	100	-200	0	-100	0	0	0	0	0	0	0	0	0	50
Total	100	50	100	100	-200	1010	-94	0	0	0	0	0	0	0	0	0	1066
Cum. +/–																	
Hardware	0	0	0	0	0	505	410	0	0	0	0	0	0	0	0	0	
Software	0	0	0	0	0	505	606	0	0	0	0	0	0	0	0	0	
Labor	100	150	250	350	150	150	50	0	0	0	0	0	0	0	0	0	
Total	100	150	250	350	150	1160	1066	0	0	0	0	0	0	0	0	0	

Chapter 11
Cost/Benefit Analysis

To perform a cost/benefit analysis the analyst must prepare two estimates. Cost projections are based on the resource requirements specifications and the future operating costs of the new system. Benefit projections are based on the goals and objectives set forth in the project contract. Although estimates of this nature are sometimes difficult to develop and justify, the analyst should prepare a rationale that explains the estimating process. This will lend some perspective to future evaluations of these estimates.

Resource Requirement Specifications

The Hardware and Software Acquisition Specifications are illustrated in Figure 11-9. These cost figures are reasonably close to the estimates that appear on the original project budget (from Figure 6-11) and are summarized below.

	Specifications	Budget
Hardware	3399	4000
Software	1095	1500
Data	***	0
People	***	***
Procedures	***	***
Miscellaneous	400	0
Tax and Shipping	500	0
	——	——
Total	5394	5500

The data resource acquisition costs are not included because this is not part of the system design at this time. However, a fax-modem is listed in the hardware and software specifications. Based on information contained in the PhonoScan brochure, data acquisition costs are estimated at $60 per month. The people and procedures costs are bundled into the labor costs associated with system documentation and training.

FIGURE 11–9 **Cornucopia Resource Requirement Specifications**

HARDWARE
Processing Platform:
Intel 486DX/33 MHz
64K Cache RAM, 4MB RAM
1.2MB and 1.44MB Drives
200MB 13ms IDE Cache Drive
ATI Graphics Ultra Video
14" CrystalScan 1024NI Color VGA Monitor
1 Parallel/2 Serial Ports
124-Key AnyKey Keyboard
Microsoft Mouse
(bundled software detailed below—#1)
Vendor: Gateway 2000 .... $2,395
Source: *PC Magazine* (9/29/92)

Peripherals:
HP DeskJet 500 414
Source: *Computer Buying World* (9/92)
Gateway 2000 TelePath Fax/Modem 195
14,400 bps data mode, 9,600 bps fax mode
(bundled software detailed below—#2)
Source: *PC Magazine* (9/29/92)
Uniscan 200 Bar Code Reader 295
Source: *InfoWorld* (9/14/92)

SOFTWARE
MS DOS 5.0 and MS Windows 3.1 (#1) 0
WinFax Pro, Crosstalk for Windows, Qmodem (#2) 0
Central Point Anti-Virus 1.2 for Windows 75
Central Point Backup 7.2 for Windows 75
The Norton Desktop 2.0 and
Prisma Your Way 2.0 for Windows 95
Vendor: Gateway 2000 (ref. #1 and #2)
Source: *PC Magazine* (9/29/92)
MS Excel for Windows (#1) 0
WordPerfect for Windows 260
Borland dBASE IV 1.5 476
Z-SOFT PC Paintbrush IV+ 1.01 114
Source: *Computer Buying World* (9/92)

DATA
None 0

PEOPLE
Training (bundled with labor cost) 0

PROCEDURES
Documentation (bundled with labor cost) 0

MISCELLANEOUS
Paper products (paper, labels) 100
Floppy disks 50
DeskJet ink supplies 100
Reference materials 75
Cleaning materials 50
Power strip surge protector 25

SUMMARY
Hardware 3,399
Software 1,095
Data 0
People 0
Procedures 0
Miscellaneous 400
=====
Subtotal 4,894
Tax & Shipping 500
=====
Total $5,394

Specific product choices have required careful consideration of several industry magazines, visits to software retailers, industry conventions, and direct calls to software manufacturers. Based on this information and the highly competitive pricing policies in the market, the analyst and user have agreed preparing a formal request for bids or proposals is not necessary.

Cost Projections

The cost projections for the delivery of the information system are identical to those on the original project budget. Even though the latest update shows that, at this time, actual expenditures are less than estimated, this is a result of the timing of hardware and software purchases and some minor inaccuracies in labor estimates. Therefore, the original budget remains unchanged.

The cost projections for the initial system implementation and continued maintenance are based on the original contract. The 12 hours of user training were projected at $25 per hour and spread over several weeks just prior to and after the system becomes operational. The maintenance costs are estimated at 1 percent per month of the $10,000 system price.

These projections are summarized below. Notice that the intangible costs are not included. Certainly the risk is very real that the system will not deliver the services it promises, or that the user will not be able to use the Sales Transaction subsystem without degrading customer service. But, because quantifying these potential costs is virtually impossible, they are simply noted and ignored in the official cost projections.

Cornucopia - Cost Projections

Cost Category	Month					
	1	2	3	4	5	6
Initial System Costs	1100	5950	1600	1150		
Training Costs				100	200	
Maintenance Costs					100	100 ->

Benefit Projections

The benefit projections are associated with the project contract (from Figure 3-10). The estimated benefits and rationale are as follows:

Decreased Reorder Costs: Contract Objective 3

> The new reordering system will allow the owner to shift this responsibility to a $5/hr. clerk, thus freeing the owner's time, which she values at $20/hr. Assuming that reordering time will be reduced to 10 to 15 hours per month, a $200 saving per month is conservative.

Increased Repeat Customers: Contract Objective 4

> The customer correspondence system will increase business, but it will take 6 months for this to show results. A conservative estimate that this will result in 10 more sales per month, at an average profit margin of $20 per sale, yields a $200 monthly benefit.

Decreased Out of Stock: Contract Objective 5

The CD reordering system will provide a more accurate record of which CDs have been sold. The owner will be able to quickly scan this report and note the items she wants the clerk to reorder. A conservative estimate is that this will reduce 5 "lost" sales per month. These benefits will begin to accrue 3 months after the new system is in place, which should be sufficient time for the most active inventory items to "turn over." At the $20 profit margin per sale, this produces $100 per month in benefits.

Cost/Benefit Analysis

Figures 11-10 and 11-11 present the cost/benefit projections in a spreadsheet and graphic format. The graph shows the payback point to be somewhere in the thirty-second month of the system. Given that the system life is estimated to be 3 to 5 years, we can reasonably assume that considerable benefits will accumulate during that period. Such benefits might be directed at further system enhancements, thus further increasing the benefits and prolonging the useful life of the system.

Cornucopia Cost/Benefit Graph **FIGURE 11–11**

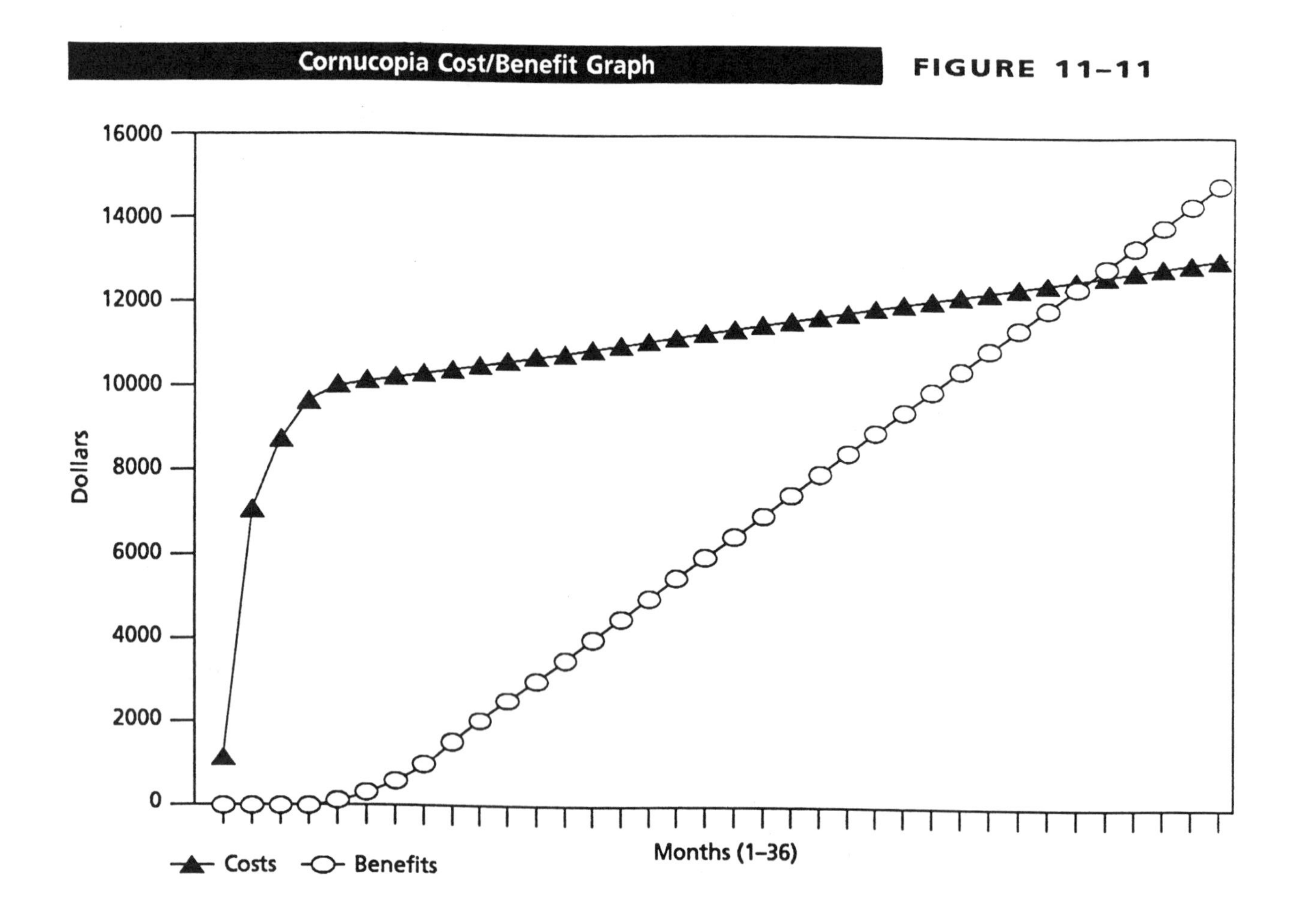

FIGURE 11–10 **Cornucopia Cost/Benefit Projections**

Date: Cost/Benefit Analysis

Month =>	1	2	3	4	5	6	7	8	9	10	11	12	13	14	15	16
Costs:																
Initial Sys.	1100	5950	1600	1150												
Training				100	200											
Maintenance					100	100	100	100	100	100	100	100	100	100	100	100
Benefits:																
Dec. Reorder Cost					200	200	200	200	200	200	200	200	200	200	200	200
Inc. Repeat Cust.										200	200	200	200	200	200	200
Dec. Out of Stock							100	100	100	100	100	100	100	100	100	100
Cumulative Costs	1100	7050	8650	9900	10200	10300	10400	10500	10600	10700	10800	10900	11000	11100	11200	11300
Cumulative Benefits	0	0	0	0	200	400	700	1000	1300	1800	2300	2800	3300	3800	4300	4800

Month =>	17	18	19	20	21	22	23	24	25	26	27	28	29	30	31	32	33	34	35	36
Maintenance	100	100	100	100	100	100	100	100	100	100	100	100	100	100	100	100	100	100	100	100
Dec. Reorder Cost	200	200	200	200	200	200	200	200	200	200	200	200	200	200	200	200	200	200	200	200
Inc. Repeat Cust.	200	200	200	200	200	200	200	200	200	200	200	200	200	200	200	200	200	200	200	200
Dec. Out of Stock	100	100	100	100	100	100	100	100	100	100	100	100	100	100	100	100	100	100	100	100
Cumulative Costs	11400	11500	11600	11700	11800	11900	12000	12100	12200	12300	12400	12500	12600	12700	12800	12900	13000	13100	13200	13300
Cumulative Benefits	5300	5800	6300	6800	7300	7800	8300	8800	9300	9800	10300	10800	11300	11800	12300	12800	13300	13800	14300	14800

Design Review Presentation

During the design review session the analyst presents several of the system design models, the most discussed of which are the menu tree and the USD. The user interfaces are demonstrated on a large-screen projection system.

The analyst points out the potential input bottleneck associated with the CD maintenance subsystem, as well as the advantages of upgrading to a point-of-sale system. The user, however, is satisfied with the design as it stands and reminds the analyst that as she understands it, these upgrades can be incorporated at a later date.

At the conclusion of the session, the analyst presents a rough schedule of activities for the development and implementation of the system. The user understands that the development phase continues the prototyping methodology used with some of the user interfaces. She also agrees to continue to participate in the project as time permits.

Time and Money

During the preceding reporting period the analyst spent 6 hours on system design, bringing this phase to 95 percent completion. One additional hour was spent on development activities, resulting in a 5 percent completion on that phase. This latter item is of some concern because it clearly identifies that the project is behind schedule. Figure 11-12 presents the Project Status Report for Week 8.

FIGURE 11–12 Cornucopia Project Status—Week 8

Date: | Cornucopia Project Status | A/O Week: 8

Activity		% Comp.	Status	1	2	3	4	5	6	7	8	9	0	1	2	3	4	5	6	Total
Analysis	-Est	100%		4	5	5	4													18
	Actuals	100%	ok	2	4	2	3													11
Design	-Est	100%					4	4	5	5	4									22
	Actuals	95%	ok			1	3	8	5	7	6									30
Develop	-Est	20%									4	4	5	5	4					22
	Actuals	5%	--								1									1
Impl.	-Est	0%													4	5	5	5	5	24
	Actuals	0%	ok																	0
Total	-Est			4	5	5	8	4	5	5	8	4	5	5	8	5	5	5	5	86
	Actuals			2	4	3	6	8	5	7	7	0	0	0	0	0	0	0	0	42
Contract		100%	ok	C																
Proj. Prelim.		100%	ok			C														
Design Rev.		100%	ok								C									
Proto. Rev.		0%	ok											S						
Train & Del.		0%	ok															S		
Final Rpt.		0%	ok																S	
				1	2	3	4	5	6	7	8	9	0	1	2	3	4	5	6	

Chapter 12
Prototyping

The reusable prototype approach is chosen for the project mainly because of the small size of the project and the straightforward nature of the general system design. The prototyped input/output and processing elements will then serve as the basis for the full system development.

System Models: Highlighted for Prototyping

To familiarize the user with prototyping and to encourage her active participation in the evaluation and revision of the prototypes, the analyst prepared the following narrative to accompany the highlighted system models discussed below.

A User's Introduction to Prototyping

The prototypes developed for this information system project are models, or facsimiles, of the real thing. After we have evaluated and revised these models, the analyst team will use them as a basis for developing the complete information system.

Each prototype was created with the understanding that revisions are more easily made during the early stages of the design-development work. Therefore, we encourage you to imagine working with the various screens and interfaces of this system and to suggest any you think may be necessary.

These models are designed to provide you with a realistic view of many of the important inputs and outputs of the new system. In some cases, you may even be able to manipulate small amounts of sample data with the prototype product. However, you should realize that models do not always behave as their real-world counterparts will. Under extreme or unusual data and processing conditions, the prototype may break down. In other words, prototypes are not substitutes for fully developed products. Even after the prototypes are perfected, a great deal of work will remain before the product is ready for use.

To help you understand exactly which parts of the information system we have prototyped, we have modified the system diagram by shading the affected input, output, and processing elements of the system.

To provide the user and the analyst with a sense of the overall prototype plan, the system flowchart (Figure 12-12) and the user's system diagram (USD) were adapted by shading the affected elements. But why were only certain portions of the system chosen for prototyping? Remembering that prototyping costs money, the analyst chose those elements that would most rapidly advance the design and allow the user to react to a computer-based product.

Cornucopia Prototype System Flowchart

FIGURE 12–12

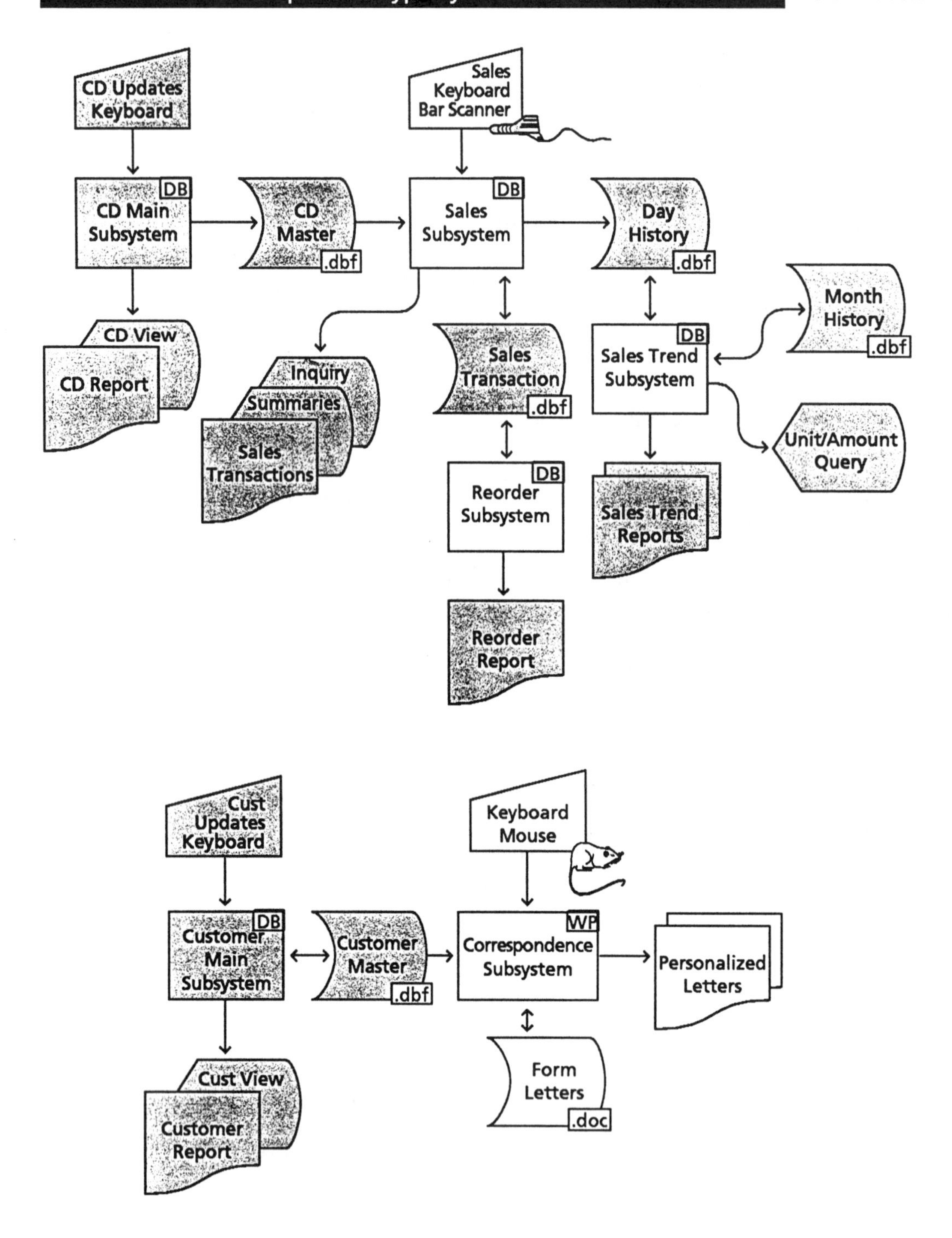

FIGURE 12–13 Cornucopia Prototype USD

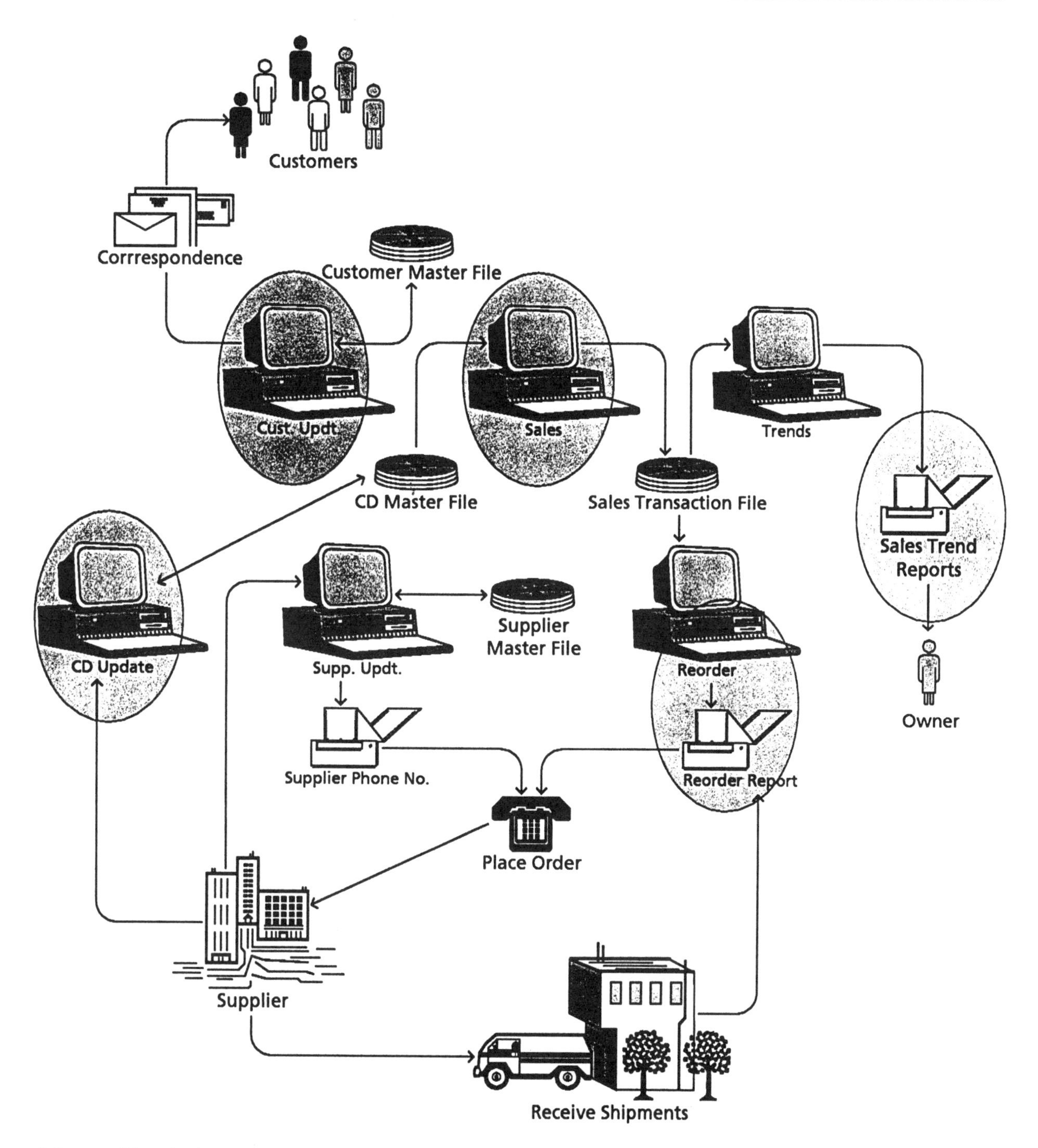

Menu Prototypes

The menu prototype can be constructed with a word processor, a paint program, or relational database program's screen painter and code builder functions. The menu trees developed during the earlier design activities (from Figures 8-15 and 9-10) were constructed with a paint program. You will recall that one purpose of these early renditions of the menu tree is to help prepare the user for the joint application design work sessions. During the prototyping activities, the menu tree serves a different purpose. Namely, it must help the user visualize how the system will work on a computer. Thus, the analyst must create a model that is easy to access and fairly compatible with the way the rest of the prototypes work. Paint program output is too unlike the database implementation the analyst envisions for the system.

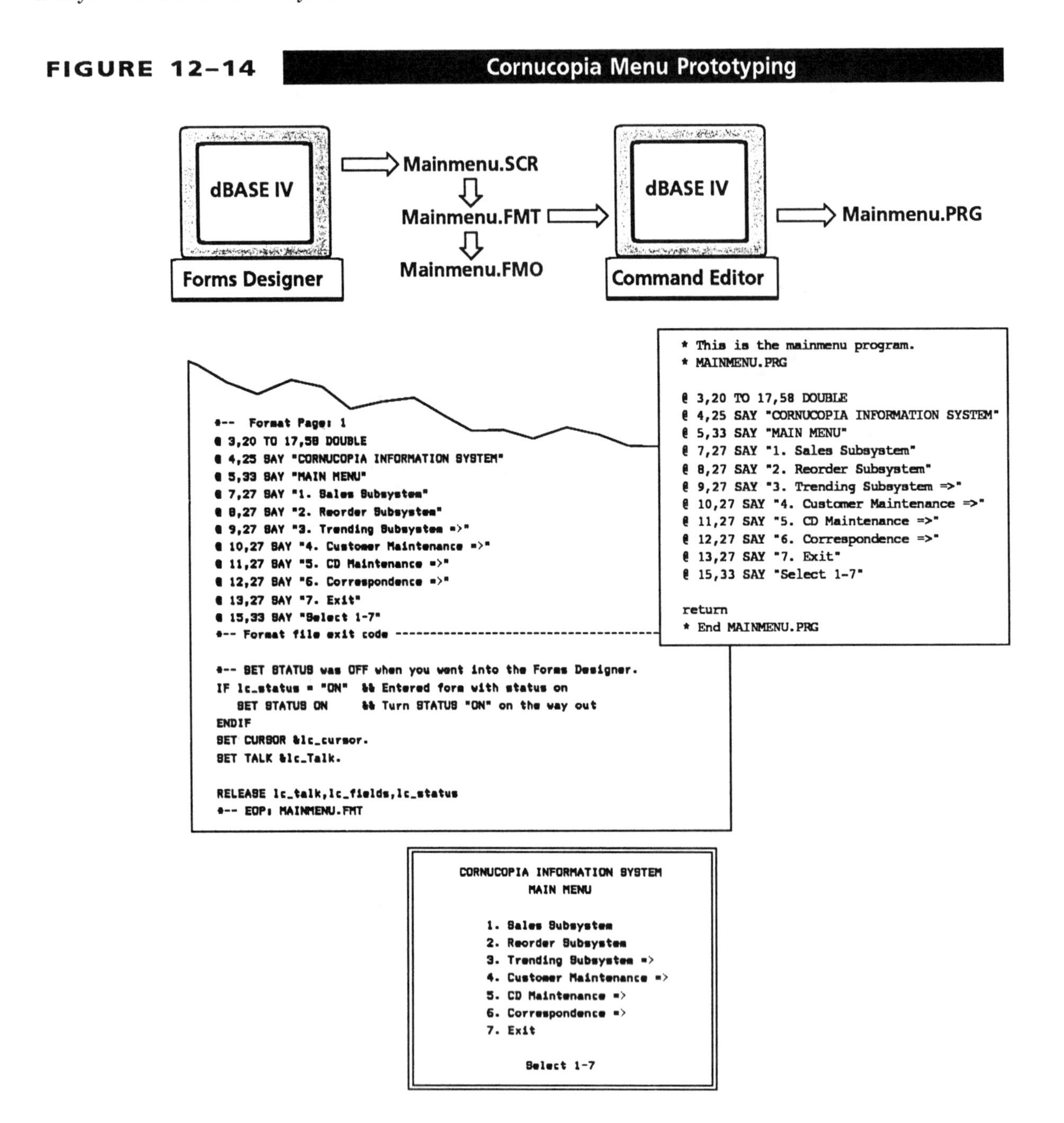

Figure 12-14 shows how the main menu prototype was created. Using dBase IV, the analyst developed the simple menu displayed at the bottom of the figure with the form screen builder. This caused dBase IV to generate three files. The .scr file is the screen file. The .fmt file is a text file containing all of the dBASE generated **source code**, which is obviously in an English-like syntax. The .fmo file contains the **machine code**, which is the same program but in a machine readable format. Of the three files, only the .fmt file can be directly manipulated by the analyst. Accordingly, the figure shows this file as input to the dBase IV command editor, which the analyst uses to remove some of the code lines and create a .prg file. The figure presents a portion of the generated source code (.fmt) and the complete, trimmed-down version of the prototype source code (.prg). The .prg file can be executed with a simple DO command, making the prototype very easy to access.

Form Prototypes

After developing the menu prototype, the analyst chose the next easiest system elements to develop: the master file maintenance forms and reports. The master file maintenance screen form sequences illustrated earlier (from Figures 8-16 and 8-17) were created from sample master files, using the forms designer and code generator feature of dBase IV. As with the menu prototype, this process yields three files. Figure 12-15 illustrates how the source code file (.fmt) is manipulated to create the prototype source code files (Custform.prg, Browcust.prg, and Blankcus.prg). The last of these files produces the blank customer maintenance screen form, illustrated previously in Figure 8-16.

You should refer to the master file maintenance subsystem structure chart (from Figure 10-18) to see exactly how the analyst plans to use these three screen forms. The modules labeled with these files' names are really nothing more than the brief code segments illustrated here.

Cornucopia Master File Form Prototyping **FIGURE 12–15**

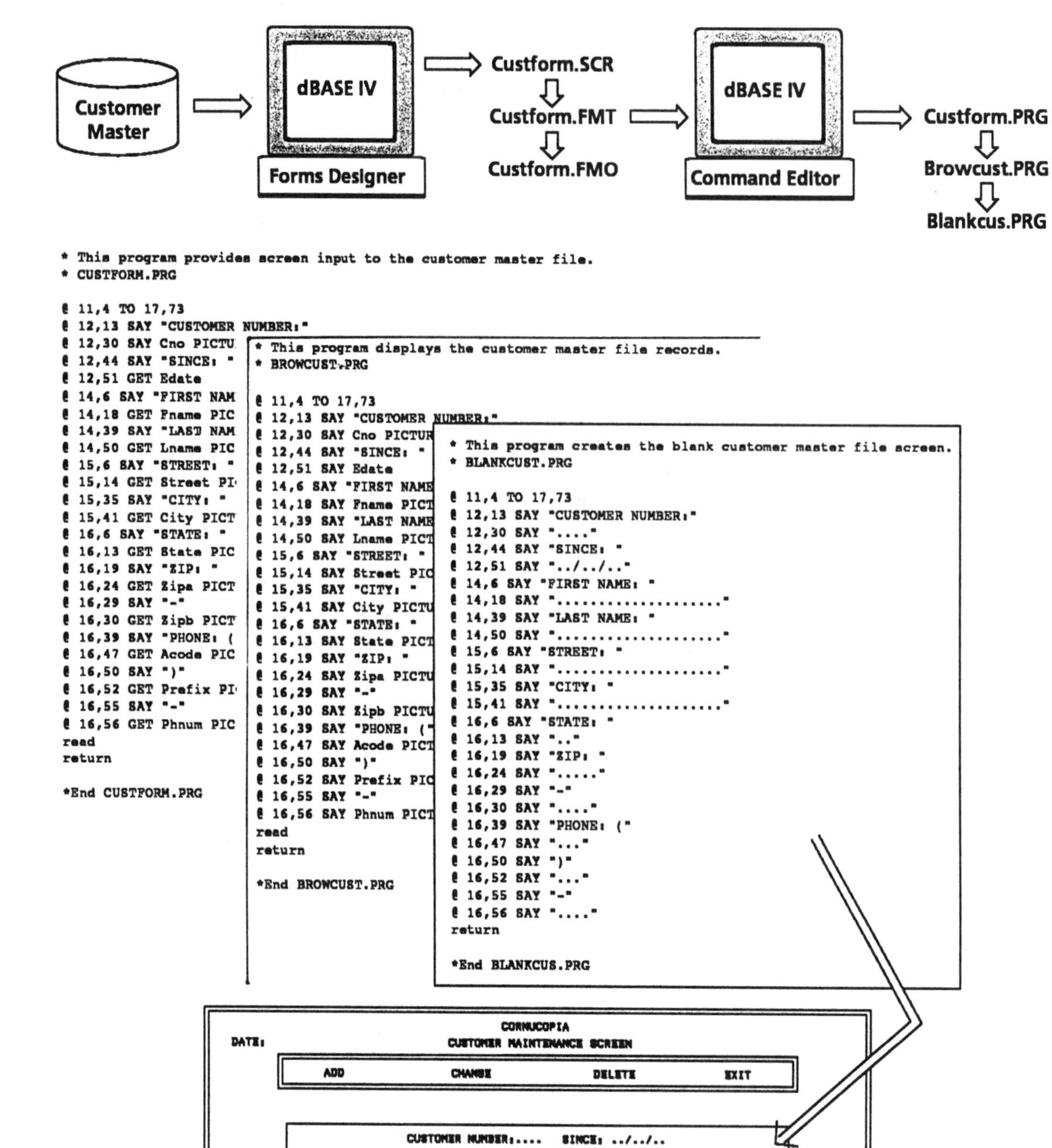

Report Prototypes

The procedure used to create the simple Customer Master File Report prototype is illustrated in Figure 12-16. In this case, however, the analyst used the reports designer, which generates the same three file types but with slightly different file extensions (.frm, .frg, and .fro). This report first appeared in the report design activities (from Figure 9-12).

FIGURE 12–16 Cornucopia Master File Report Prototyping

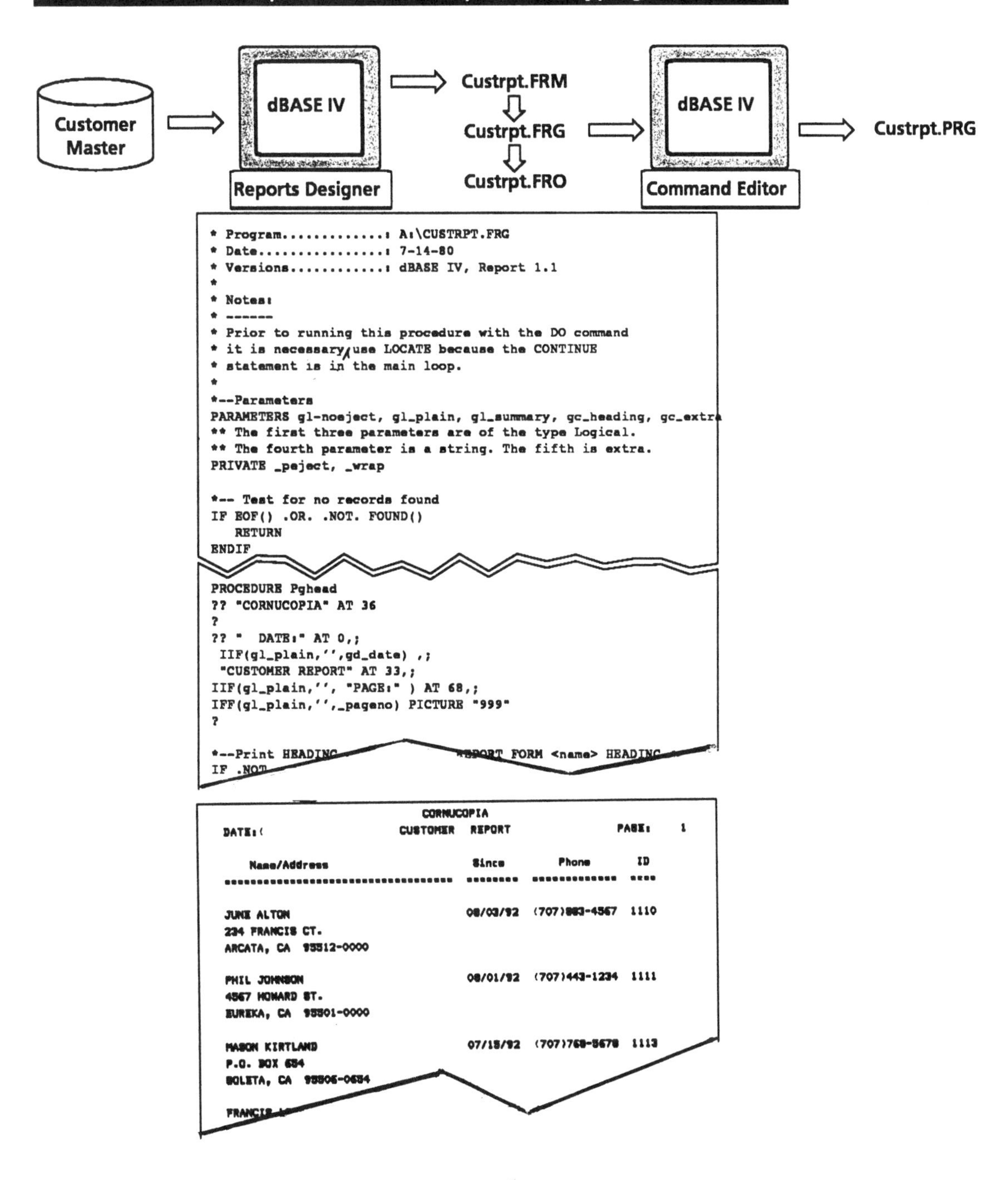


```
* Program.............: A:\CUSTRPT.FRG
* Date................: 7-14-80
* Versions............: dBASE IV, Report 1.1
*
* Notes:
* ------
* Prior to running this procedure with the DO command
* it is necessary use LOCATE because the CONTINUE
* statement is in the main loop.
*
*--Parameters
PARAMETERS gl-noeject, gl_plain, gl_summary, gc_heading, gc_extr
** The first three parameters are of the type Logical.
** The fourth parameter is a string. The fifth is extra.
PRIVATE _peject, _wrap

*-- Test for no records found
IF EOF() .OR. .NOT. FOUND()
   RETURN
ENDIF

PROCEDURE Pghead
?? "CORNUCOPIA" AT 36
?
?? " DATE:" AT 0,;
 IIF(gl_plain,'',gd_date) ,;
 "CUSTOMER REPORT" AT 33,;
IIF(gl_plain,'', "PAGE:" ) AT 68,;
IFF(gl_plain,'',_pageno) PICTURE "999"
?

*--Print HEADING          REPORT FORM <name> HEADING
IF .NOT
```

```
                          CORNUCOPIA
DATE:(                 CUSTOMER  REPORT                PAGE:   1

   Name/Address                      Since     Phone       ID
================================= ======== ============= ====

JUNE ALTON                        08/03/92 (707)883-4567 1110
234 FRANCIS CT.
ARCATA, CA  95512-0000

PHIL JOHNSON                      08/01/92 (707)443-1234 1111
4567 HOWARD ST.
EUREKA, CA  95501-0000

MASON KIRTLAND                    07/15/92 (707)768-5678 1113
P.O. BOX 654
GOLETA, CA  95506-0654

FRANCIS
```

Sales Transaction Prototypes

The Sales Transaction subsystem posed a more formidable challenge to the analyst. However, because this subsystem is at the heart of the project design, including it in the prototype seemed appropriate. After creating a sample of the sales transaction file, the analyst used the forms designer and a reiteration of the preceding process to create the three screen forms previously illustrated in Figure 8-18. The Daily Sales Summary output prototype (Figure 9-13, item 4) followed in a similar fashion.

With the exception of item 11, all the output prototypes that appear in Figures 9-14 and 9-15 were created using the reports designer, along with sample history files (Dayhist and Mohist). Figure 9-14, item 11 and the outputs in Figure 9-16 were created with a paint program, so there is no code behind these prototypes.

Time and Money

The analyst spent 8 hours transforming the design concepts into sample files and prototypes, which is reported under the development category on the project status report. This brings the estimate of completion up to 20 percent, which is still well below what the analyst had expected. It also appears that the labor hours saved during the analysis phase have been more than consumed during the design phase, with the prospect of more overruns during the development phase. When asked why this is so, the analyst reported that although no problems occurred while creating the master file prototypes, the sales transaction processing prototype was more difficult because of the complex file relationships involved.

The project budget for Week 9 is presented in Figure 12-17. Notice that the expenditures for hardware and software seem to be complete and well within the estimates established for each category. On the other hand, the labor charges exceed the estimates.

The analyst should carefully note that the computer hardware and software performance-price relationship is volatile. Based on what is currently available, one of two things should happen to your project budgets and resource purchases. Either the budgets should be reduced or the products should be upgraded. For example, today $2,395 will buy more than the 486DX-33 itemized on the resource requirements specifications (from Figure 11-9). In this case, rather than upgrade the hardware, the analyst chose to save the $400 difference between the actual purchase price ($1,995) and the budgeted amount ($2,395). Also, software prices have gone down since the budget was developed. These savings more than offset the labor charge excesses noted earlier. However, the question remains: To what extent is the analyst free to cover labor cost overruns with hardware and software cost savings?

FIGURE 12–17 **Cornucopia Project Budget—Week 9**

Date: Cornucopia Budget A/O Wk: 9

	Week1	Week2	Week3	Week4	Week5	Week6	Week7	Week8	Week9	Week10	Week11	Week12	Week13	Week14	Week15	Week16	Total
Estimates																	
Hardware						2500	500	500	500								4000
Software						1000	250	250									1500
Labor	200	250	250	400	200	250	250	400	200	250	250	400	250	250	250	250	4300
Total	200	250	250	400	200	3750	1000	1150	700	250	250	400	250	250	250	250	9800
Actuals																	
Hardware						1995	595	495									3085
Software						495	149	289	235								1168
Labor	100	200	150	300	400	250	350	350	400								2500
Total	100	200	150	300	400	2740	1094	1134	635	0	0	0	0	0	0	0	6753
Weekly +/–																	
Hardware	0	0	0	0	0	505	-95	5	500	0	0	0	0	0	0	0	915
Software	0	0	0	0	0	505	101	-39	-235	0	0	0	0	0	0	0	332
Labor	100	50	100	100	-200	0	-100	50	-200	0	0	0	0	0	0	0	-100
Total	100	50	100	100	-200	1010	-94	16	65	0	0	0	0	0	0	0	1147
Cum. +/–																	
Hardware	0	0	0	0	0	505	410	415	915	0	0	0	0	0	0	0	
Software	0	0	0	0	0	505	606	567	332	0	0	0	0	0	0	0	
Labor	100	150	250	350	150	150	50	100	-100	0	0	0	0	0	0	0	
Total	100	150	250	350	150	1160	1066	1082	1147	0	0	0	0	0	0	0	

Chapter 13
4GL Programming

Although the analysts are aware of the newer object-oriented horizontal software, they decide to develop the majority of this project with a straightforward relational database product. A considerable amount of code accompanies the following illustrations. For the most part, the dBASE IV code was hand-coded, although the Application Generator utility could have created much of it automatically. All of the WordPerfect mail merge and Quattro Pro macro instructions were generated using a Macro Record utility.

Prototype Conversion

The menu, master file maintenance screen forms, and sales transaction processing output prototypes serve as the basis for developing the working product. The following itemizes the prototypes and their associated processing programs that are involved in this process:

Prototype	**Associated Process**
Main System Menu	To be determined
Customer Master File Maintenance Screen Forms	Customer Maintenance (custmain.prg)
CD Master File File Maintenance Screen Forms	CD Maintenance (cdmain.prg)
Sales Transaction File Screen Forms	Sales Processing (saletran.prg)
Master File Reports	To be determined
Sales Transaction File Reports	To be determined
Sales Queries & Inquires	To be determined
Sales Trend Reports	To be determined
CD Reorder Report	To be determined

The "To be determined" associated program notation for some of the prototypes indicates that the prototype was created with a very primitive program or screen painting utility. Therefore, these programs still need to be generated or hand constructed.

The three programs specifically associated with the prototypes were initially created with the dBASE IV form-design utility. After much of the documentary and system environment code was removed from these programs, they were modified as noted in the following.

Application Development Strategy

After much debate, the analysts and user agreed that the information system should be directly accessible from Windows. Further, it was agreed that from this entry point the user should be able to initiate the database application that implements the bulk of the processing in this information system. The development of the Windows linkage is postponed to the next chapter. Developing the main database menu, on the other hand, is a task that must accompany the development of the different database subsystems.

Figure 13-11 illustrates the main menu screens developed to provide the user with quick access to five subsystems: CD Maintenance, Customer Maintenance, Sales, Sales Trends and CD Reordering. This interface design is patterned after the dBASE IV interface to minimize the confusion that might arise when the user refers to Borland's dBASE IV documentation or when the user chooses to investigate the dBASE IV program independently. Notice that the Reorder subsystem is "inactive." This demonstrates an important developmental strategy in which each subsystem is developed and tested separately. The menuing system is constructed with such "inactive" notations and later modified as working subsystems are incorporated into the product.

The implementing dBASE IV code for these menu screens is presented in Figure 13-12. Although this code may look formidable to the novice, a careful inspection of the detail reveals a great deal of repetition, along with some readily identifiable nonprocedural statements. For example, the define command is used to specify what the programmer wants the menu to look like without actually detailing how this is to be accomplished. Further, this program could have been generated by the dBASE IV Application Generator, eliminating the hand coding altogether.

Master File Maintenance Development

The design for the Customer master file maintenance subsystem appears in Figure 13-13. This design involves a controlling program (custmain.prg), three primary processing modules (custadd.prg, custchg.prg, custdel.prg), and several submodules (custform.prg,blankcus.prg, etc.). The structure chart shows exactly how these small programs work together to present the sequence of menus and screen forms that permit the user to maintain the master file.

The implementing code is presented in Figures 13-14, 13-15 and 13-16. Without question, this code is formidable. To make matters worse, the custom screen sequences used in this maintenance subsystem preclude the use of the Application Generator. Therefore, no substitute exists for hand coding most of these programs. The good news is that the CD Maintenance subsystem is almost identical to the Customer Maintenance subsystem. So, once the analyst-programmer has perfected one subsystem, creating the other is fairly easy.

FIGURE 13–11 Cornucopia Database Applications Main Menu Screens

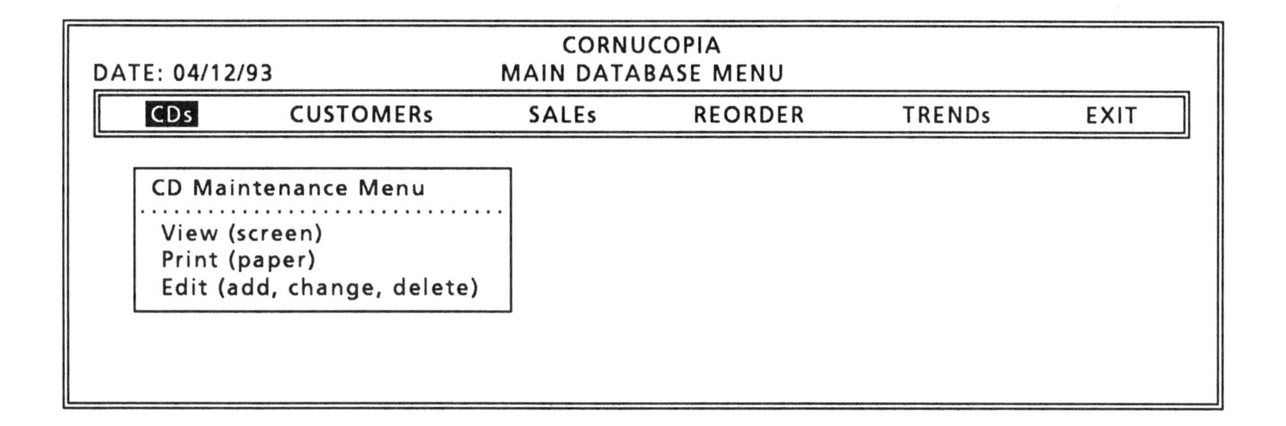

CORNUCOPIA
DATE: 04/12/93
MAIN DATABASE MENU

CDs CUSTOMERs **SALEs** REORDER TRENDs EXIT

Sales Transaction Subsystem

Press ENTER to activate.

CORNUCOPIA
DATE: 04/12/93
MAIN DATABASE MENU

CDs CUSTOMERs SALEs **REORDER** TRENDs EXIT

CD Reorder Subsystem

Subsystem INACTIVE

Cornucopia MAINMENU.PRG Program Listing

FIGURE 13–12

```
* This is the main menu program.
* MAINMENU.PRG
clear
clear all
set talk off
set status off
set scoreboard off

set default to D:
set directory to D:\

@ 1.1 to 21,77 double
@ 2,35 say "CORNUCOPIA"
@ 3,4 say "DATE:"
@ 3,9 say date()
@ 3,31 say "MAIN DATABASE MENU"
define menu mmain
        define pad mp1 of mmain prompt "CDs" at 5,9
        define pad mp2 of mmain prompt "CUSTOMERs" at 5,16
        define pad mp3 of mmain prompt "SALEs" at 5,30
        define pad mp4 of mmain prompt "REORDER" at 5,40
        define pad mp5 of mmain prompt "TRENDs" at 5,53
        define pad mp6 of mmain prompt "EXIT" at 5,65

        on pad mp1 of mmain activate popup cdpop
        on pad mp2 of mmain activate popup custpop
        on pad mp3 of mmain activate popup salepop
        on pad mp4 of mmain activate popup reordpop
        on pad mp5 of mmain activate popup trendpop
        on selection pad mp6 of mmain deactivate menu
define popup cdpop from 8,7 to 14,37
        define bar 1 of cdpop prompt " CD Maintenance Menu" skip
        define bar 2 of cdpop prompt "..............................." skip
        define bar 3 of cdpop prompt " View (screen)"
        define bar 4 of cdpop prompt " Print (paper)"
        define bar 5 of cdpop prompt " Edit (add, change, delete)"
        on selection popup cdpop do  \cdmain\cdmain.prg
define popup custpop from 8,14 to 14,44
        define bar 1 of custpop prompt " Customer Maintenance Menu" skip
        define bar 2 of custpop prompt "............................" skip
        define bar 3 of custpop prompt " View (screen)"
        define bar 4 of custpop prompt " Print (paper)"
        define bar 5 of custpop prompt " Edit (add, change, delete)"
        on selection popup custpop do  \custmain\custmain.prg
define popup salepop from 8,28 to 14,58
        define bar 1 of salepop prompt " Sales Transaction Subsystem" skip
        define bar 2 of salepop prompt "............................" skip
        define bar 3 of salepop prompt " Press ENTER to activate."
        on selection popup salepop do  \saletran\saletran.prg
define popup reordpop from 8,38 to 14,68
        define bar 1 of reordpop prompt "CD Reorder Subsystem" skip
        define bar 2 of reordpop prompt "..........................." skip
        define bar 3 of reordpop prompt "Subystem INACTIVE" skip
define popup reordpop from 8,43 to 14,73
        define bar 1 of reordpop prompt " Sales Trending Subsystem" skip
        define bar 2 of reordpop prompt "..........................." skip
        define bar 3 of reordpop prompt " Subystem INACTIVE" skip
@ 4,4 to 6,74 double
activate menu mmain
release popup cdpop
release popup custpop
release popup salepop
release popup reordpop
release popup trendpop
release menu mmain
clear
clear all
set talk on
set status on
set scoreboard on
quit
* End MAINMENU.PRG
```

FIGURE 13–13 **Cornucopia Structure Chart for Customer Maintenance Subsystem**

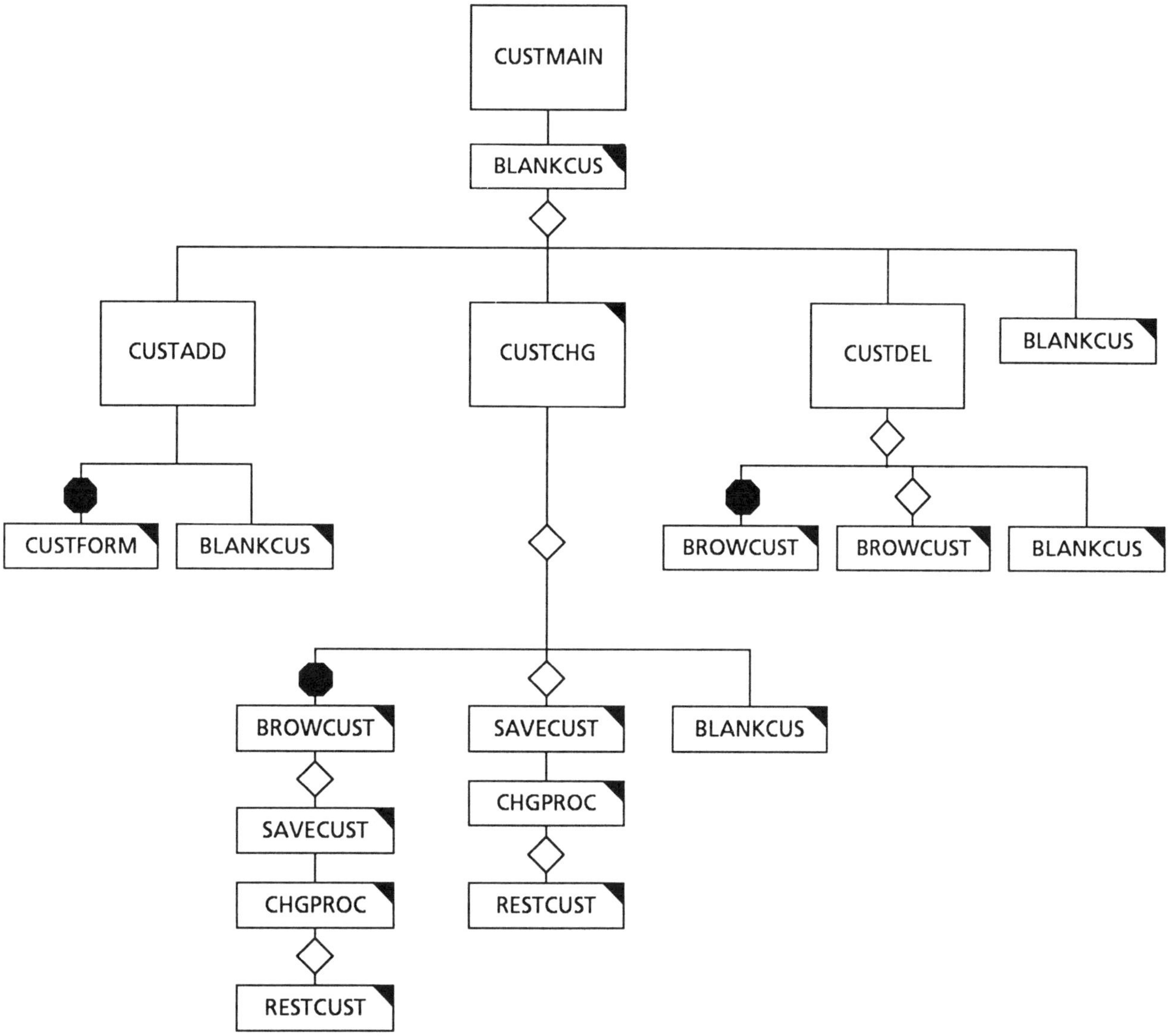

CUSTMAIN
BLANKCUS
CUSTADD
CUSTCHG
CUSTDEL
BLANKCUS
CUSTFORM
BLANKCUS
BROWCUST
BROWCUST
BLANKCUS
BROWCUST
SAVECUST
BLANKCUS
SAVECUST
CHGPROC
CHGPROC
RESTCUST
RESTCUST

Cornucopia CUSTMAIN.PRG and CUSTADD.PRG Program Listings

FIGURE 13–14

```
* This is the customer maintenance menu program.
* CUSTMAIN.PRG

clear
@ 1,1 to 21,77 double
@ 2,35 say "CORNUCOPIA"
@ 3,4 say "DATE:"
@ 3,9 say date()
@ 3,27 say "CUSTOMER MAINTENANCE SCREEN"

set directory to \custmain
use customer.dbf
public mustpack
mustpack = .f.

define menu cmain
        define pad p1 of cmain prompt "ADD" at 5,10
        define pad p2 of cmain prompt "CHANGE" at 5,20
        define pad p3 of cmain prompt "DELETE" at 5,37
        define pad p4 of cmain prompt "EXIT" at 5,55

        on selection pad p1 of cmain do custadd.prg
        on pad p2 of cmain active popup cchange
        on pad p3 of cmain activate popup cdelete
        on selection pad p4 of cmain deactivate menu

define popup cchange from 5,27 to 8,35
        define bar 1 of cchange prompt "Browse"
        define bar 2 of cchange prompt "Select"
        on selection popup cchange do custchg.prg

define popup cdelete from 5,44 to 8,52
        define bar 1 of cdelete prompt "Browse"
        define bar 2 of cdelete prompt "Select"
        on selection popup cdelete do custdel.prg

do blankcus.prg
@ 4,4 to 6,74 double
activate menu cmain

release popup cchange
release popup cdelete
release menu cmain

if mustpack = .t.
   @ 5,10 say 'Please wait - packing the customer master file.'
   goto top
   set filter to cno = 0000
   delete all
   pack
endif

@ 3,27 clear to 3,67
@ 3,31 say "MAIN DATABASE MENU"
@ 5,8 clear to 5,67
@ 5,9 say "CDs"
@ 5,16 say "CUSTOMERs"
@ 5,30 say "SALEs"
@ 5,40 say "REORDER"
@ 5,53 say "TRENDs"

@ 5,65 SAY "EXIT"
@ 11,4 clear to 17,73

close databases
set directory to \
return

* End CUSTMAIN.PRG
```

```
* This is the customer master file ADD program
* CUSTADD.PRG

set order to cno
contadd = 'Y'
do while contadd = 'Y'
        goto bottom
        newcno = cno + 1
        append blank
        replace cno with newcno
        replace status with 'A'
        action = 'C'
        do while action = 'C'
           do custform.prg
           @ 18,15 say 'Action: U(pdate),C(hange screen),A(bort)'
           @ 18,57 get action picture "@! X"
           read
        enddo
        if action = 'A'
           mustpack = .t.
           replace cno with 0000
           replace status with 'D'
        endif
        @ 19,23 say 'Add another customer? (Y/N)'
        @ 19,51 get contadd picture "@! X"
        read
enddo
do blankcus.prg
@ 18,15 clear to 19,57
return

* End CUSTADD.PRG
```

FIGURE 13–15 Cornucopia CUSTCHG.PRG and SAVECUST.PRG Program Listings

```
* This is the customer master file CHANGE program.
* CUSTCHG.PRG

set filter to status = 'A'
if bar() = 1
   set order to lname
   goto top
   action = 'B'
   do while action = 'B' .and. .not. eof()
      do browcust.prg
      action = 'B'
      @ 18,15 say 'ACTION: C(hange),B(rowse),A(bort)'
      @ 18,49 get action picture "@! X"
      read
      if action = 'B'
         skip
      endif
   enddo
   if action = 'C'
      do savecust.prg
      do chgproc
   endif
endif

if bar() = 2
  set order to cno
  goto top
  cnoreq = 0000
  @ 9,15 say 'Enter the customer account number:'
  @ 9,50 get cnoreq picture "9999"
  read
  seek cnoreq
  if found ()
      @ 10,15 say 'Proceed with your changes.'
      do savecust.prg
      do chgproc
  else
      @ 10,15 say 'This customer number is not on file.'
      timer = 0
      do while timer < 5000
         timer = timer + 1
      enddo
  endif
  @ 9,15 clear to 10,53
endif

do blankcus.prg
@ 18,15 clear to 19,57
return

procedure chgproc
   action = 'C'
   do while action = 'C'
      do custform.prg
      @ 18,15 say 'ACTION: U(pdate),C(hange screen),A(bort)'
      @ 18,57 get action picture "@! X"
      read
   enddo
   if action = 'A'
      do restcust.prg
   endif
return

* End CUSTCHG.PRG
```

```
* The program stores the customer master file record.
* SAVECUST.PRG

public slname,sfname,sstreet,scity,sstate,szipa,szipb
public sacode,sprefix,sphnum,sedate,sstatus

slname = lname
sfname = fname
sstreet = street
scity = city
sstate = state
szipa = zipa
szipb = zipb
sacode = acode
sprefix = prefix
sphnum = phnum
sedate = edate
sstatus = status
return

* End SAVECUST.PRG
```

Cornucopia CUSTDEL.PRG and RESTCUST.PRG Program Listings FIGURE 13–16

```
* This is the customer master file DELETE program.
* CUSTDEL.PRG

set filter to status = 'A'
if bar() = 1
   set order to lname
   goto top
   action = 'B'
   do while action = 'B' .and. .not. eof()
      do browcust.prg
      action = 'B'
      @ 18,15 say 'ACTION: D(elete),B(rowse),A(bort)'
      @ 18,49 get action picture "@! X"
      read
      if action = 'B'
         skip
      endif
   enddo
  if action = 'D'
      replace status with 'D'
  endif
endif

if bar() = 2
  set order to cno
  goto top
  cnoreq = 0000
  @ 9,15 say 'Enter the customer account number:'
  @ 9,50 get cnoreq picture '9999'
  read
  seek cnoreq
  if found()
      @ 10,15 say 'Proceed with your delete.'
      do browcust.prg
      action = 'D'
      @ 18,15 say 'ACTION: D(elete),A(bort)'
      @ 18,40 get action picture "@! X"
      read
      if action = 'D'
         replace status with 'D'
      endif
  else
      @ 10,15 say 'This customer number is not on file.'
      timer = 0
      do while timer < 5000
         timer = timer + 1
      enddo
  endif
  @ 9,15 clear to 10,53
endif

do blankcus.prg
@ 18,15 clear to 19,57
return

* End CUSTDEL.PRG
```

```
* This program restores the customer master file record.
* RESTCUST.PRG

replace lname with slname
replace fname with sfname
replace street with sstreet
replace city with scity
replace state with sstate
replace zipa with szipa
replace zipb with szipb
replace acode with sacode
replace prefix with sprefix
replace phnum with sphnum
replace edate with sedate
replace status with sstatus
return

* End SAVECUST.PRG
```

Sales Transaction Processing Development

The structure chart for the Sales Transaction subsystem appears in Figure 13-17. Three main modules to this process are:sales.prg,salesum.prg, and upsales.prg. The Sales module is responsible for capturing the transaction data and adding it to the Sale database file. The Salesum module generates the Daily Sales Summary screen report (from Figure 9-13, item 4). The Upsales module writes a summary record to the Dayhist database file.

FIGURE 13–17 Cornucopia Sales Subsystem Structure Chart Revisited

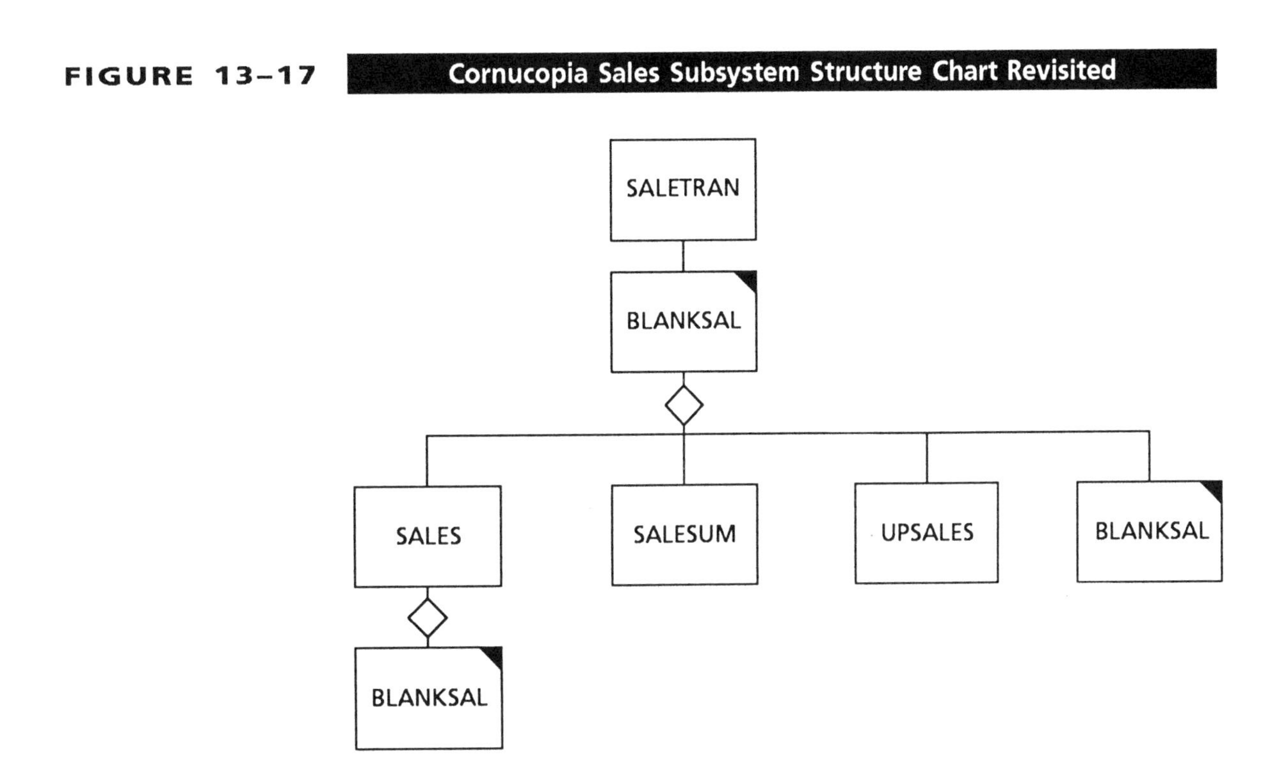

The code for this subsystem appears in Figures 13-18, 13-19 and 13-20. Once again, it is complex. By now, however, you should recognize that dBASE IV code employs a combination of procedural and nonprocedural language techniques. As lengthy and convoluted as these modules may seem, they are much less so than if a third-generation language were used. An experienced programmer will have little difficulty walking through this logic.

But, what about the analyst who has a limited programming experience? Some would suggest that visual, object-oriented programming software holds the answer. This remains to be seen. In the meantime, an analyst with this background will need to team up with someone who does have the skills required to link such programs together or at least to modify existing programs.

Cornucopia SALETRAN.PRG and BLANKSAL.PRG Program Listings

FIGURE 13–18

```
* This is the sales transaction menu program.
* SALETRAN.PRG

clear
@ 1,1 to 21,77 double
@ 2,34 say "CORNUCOPIA"
@ 3,4 say "DATE:"
@ 3,9 say date()
@ 3,28 say "SALES TRANSACTION SCREEN"

set directory to \saletran
select 1
   use \cdmain\cd.dbf
select 2
   use sales.dbf
select 3
   use dayhist.dbf

select 1
public entries
entries = 0

define menu tmain
        define pad p1 of tmain prompt "ENTER SALE" at 5,13
        define pad p2 of tmain prompt "DAILY SUMMARY" at 5,30
        define pad p3 of tmain prompt "UPDATE" at 5,50
        define pad p4 of tmain prompt "EXIT" at 5,61

        on selection pad p1 of tmain do sales.prg
        on selection pad p2 of tmain do salesum.prg
        on selection pad p3 of tmain do upsales.prg
        on selection pad p4 of tmain deactivate menu

do blanksal.prg
@ 4,4 to 6 74 double
activate menu tmain

release menu tmain

@ 3,28 clear to 3,67
@ 3,31 say "MAIN DATABASE MENU"
@ 5,8 clear to 5,67
@ 5,9 say "CDs"
@ 5,16 say "CUSTOMERs"
@ 5,30 say "SALEs"
@ 5,40 say "REORDER"
@ 5,53 say "TRENDS"
@ 5,65 say "EXIT"
@ 7,2 clear to 20,76

close databases
set directory to \
return

* End SALETRAN.PRG
```

```
* This program displays a blank sales transaction screen data area.
* BLANKSAL.PRG

@ 8,5 TO 19,74
@ 9,9 SAY "CD NUMBER"
@ 9,29 SAY "TITLE/ARTIST"
@ 9,55 SAY "LIST"
@ 9,63 SAY "SALE"
@ 9,70 SAY "OK"
@ 10,8 SAY "============"
@ 10,22 SAY "=============================="
@ 10,54 SAY "======"
@ 10,70 SAY "=="
@ 11,8 SAY "............"
@ 11,22 SAY ".............................."
@ 11,54 SAY "......"
@ 11,62 SAY "......"
@ 11,70 SAY ".."
@ 12,22 SAY ".............................."
@ 14,8 SAY "............"
@ 14,22 SAY ".............................."
@ 14,54 SAY "......"
@ 14,62 SAY "......"
@ 14,70 SAY ".."
@ 15,22 SAY ".............................."
@ 17,8 SAY "............"
@ 17,22 SAY ".............................."
@ 17,54 SAY "......"
@ 17,62 SAY "......"
@ 17,70 SAY ".."
@ 18,22 SAY ".............................."
return

* End BLANKSAL.PRG
```

FIGURE 13–19 Cornucopia SALES.PRG Program Listing

```
* This is the sales transaction capture program.
* SALES.PRG

if entries = 3
    do blanksal.prg
    entries = 0
endif
do case
    case entries = 0
        row1 = 11
        row2 = 12
    case entries = 1
        row1 = 14
        row2 = 15
    case entries = 2
        row1 = 17
        row2 = 18
endcase

set order to cdno
goto top
action = 'C'
do while action = 'C'
   @ 7,10 clear to 7,65
   salecdno = space(12)
   @ 7,10 say "Enter the CD number."
   @ row1,8 get salecdno picture "999999999999"
   read
   if salecdno = space(12)
      salecdno = "000000000000"
   endif

   seek salecdno
   if found()
      @ 7,10 clear to 7,65
      @ row1,22 say title
      @ row1,54 say lprice
      @ row2,22 say artist
      override = lprice
      @ 7,25 say "Override list price or <Enter>."
      @ row1,62 get override picture "999.99"
      read
   else
      @ 7,31 say "This CD is not on the master file."
      timer = 0
      do while timer < 5000
         timer = timer + 1
      enddo
      @ 7,10 clear to 7,65
      @ row1,22 say "Transaction aborted"
      action = 'A'
      exit
   endif

   action = 'P'
   @ 7,25 say "OK to process this transaction?"
   @ 20,25 say "ACTION: P(rocess),A(bort)"
   @ 20,51 get action picture "@! X"
   read
   if action = 'P'
      select 2
      append blank
      replace sdate with date()
      replace cdno with salecdno
      replace sprice with override
      replace sposted with .f.
      @ row1,70 say "OK"
      select 1
   endif
enddo

if action = 'A'
   @ row1,22 say "..............................."
   @ row1,54 say "......"
   @ row1,62 say "......"
   @ row2,22 say "..............................."
   @ row1,22 say "Transaction aborted."
endif

@ 7,10 clear to 7,65
@ 20,25 clear to 20,55
entries = entries + 1
return

* End SALES.PRG
```

Cornucopia UPSALES.PRG and SALESUM.PRG Program Listings

FIGURE 13–20

```
* This program summarizes the daily sales to the DAYHIST file.
* UPSALES.PRG

select 2
set order to sdate
goto top
set filter to sposted = .f.
seek date()

units1 = 0
units2 = 0
units3 = 0
amt1 = 0
amt2 = 0
amt3 = 0

do while sdate = date()
   do case
      case sprice > 20.00
          units3 = units3 + 1
          amt3 = amt3 + sprice
      case sprice > 11.00
          units2 = units2 + 1
          amt2 = amt2 + sprice
      otherwise
          units1 = units1 + 1
          amt1 = amt1 + sprice
   endcase
   replace sposted with .t.
   skip
enddo

select 3
set order to hdate
seek date()
if found()
   replace dayunits1 with dayunits1 + units1
   replace dayunits2 with dayunits2 + units2
   replace dayunits3 with dayunits3 + units3
   replace dayamt1 with dayamt1 + amt1
   replace dayamt2 with dayamt2 + amt2
   replace dayamt3 with dayamt3 + amt3
else
   append blank
   replace hdate with date()
   replace dayunits1 with units1
   replace dayunits2 with units2
   replace dayunits3 with units3
   replace dayamt1 with amt1
   replace dayamt2 with amt2
   replace dayamt3 with amt3
   replace dposted with .f.
endif

select 1
return

* End UPSALES.PRG
```

```
* This program displays sales summaries for the day.
* SALESSUM.PRG

select 2
set order to sdate
goto top
seek date()

unitsold = 0
amtsold = 0.0
do while sdate = date()
   unitsold = unitsold + 1
   amtsold = amtsold + sprice
   skip
enddo

define window summary from 7,5 to 19,74
activate window summary
   @ 2,25 say "DAILY SALES SUMMARY"
   @ 2,53 say time()
   @ 3,24 say "====================="
   @ 5,25 say "UNITS:"
   @ 5,32 say unitsold picture "9999"
   @ 7,25 say "AMOUNT:"
   @ 7,34 say amtsold picture "99,999.99"
   @ 10,0
   wait
deactivate window summary

select 1
return

* End SALESUM.PRG
```

Correspondence Subsystem

The Correspondence subsystem was identified at the very beginning of the project, but this is the first serious mention of exactly how the analysts intend to satisfy this information requirement. As depicted in the system flowchart (from Figure 10-17), the Customer Master File Maintenance subsystem plays a crucial role in the Correspondence subsystem. The other major component is the file of form letters developed with a word processing program. In combination, these two elements form the basis for a mail-merge operation, in which form letters are "personalized" for selected records in the Customer master file.

Figure 13-21 presents the system flowchart, the macro menu and the macro code for a mail-merge operation. The system flowchart shows two input files to the mail-merge macro mrgmac1.wcm, one of which is a form letter. The other file is the Customer master file, transformed into a file that can be imported by the word processor. The output from this operation is a file of letters (mrglet.doc), each personally tailored to a different customer. The WordPerfect Macro menu option shows that the user can easily initiate this particular mail-merge operation by selecting Generate Letter 1. Finally, the macro code appears as what may strike you as a strange collection of commands, some with familiar word processing actions (FileOpen, SearchReplace, etc.) and some with odd syntax (ReplacementScope:Extended).

Fortunately, all of this code was generated by WordPerfect via the macro recorder function. Unfortunately, this version of WordPerfect for Windows is not very "import friendly" to dBASE IV database files. Notice in Figure 13-22 the two-step operation required to transform the customer database to an ASCII delimited file and then to a secondary file with the quotation marks removed. This illustrates another important reason why the analyst should carefully investigate software compatibility during the system resource specifications activities.

Figure 13-23 presents a sample of each type of letter. The primary file uses conditional logic to distinguish between "active" and "inactive" customers, thereby allowing notices of special offers to "preferred customers."

Spreadsheet Application

Although not specified in the system design, the user and analyst agree that a report on the relative profitability of sales in the different price ranges should be included in the system. Figures 13-24, 13-25 and 13-26 present the steps required to generate a Sales/Profit Chart that satisfies this need. This report summarizes the sale and profit amounts and percentages by price category. It also allows the user to specify the number of days to be included in the summary.

The system flowchart in Figure 13-24 shows the transformation of the Dayhist database file into an ASCII delimited file, which is then imported into the spreadsheet program. The saless1.wb1 file contains the spreadsheet macro that builds the Sales/Profit Chart. The macro code is presented in Figure 13-25. Although some elements of the code were first created through the macro recorder, this macro was hand coded for the most part. Notice that some elements of modular design are evident even in this small macro. The commands located to the right of the main body of macro code (_moveup, _movedown, etc.) are referred to, or called from, the macro. One real advantage to this technique is that it makes the macro easier to understand.

Figure 13-26 illustrates the cell formulas and the resulting Quattro Pro for Windows screen output. When the user presses any key, the macro ends and the spreadsheet program is closed.

Cornucopia Correspondence: Mail Merge Macro

FIGURE 13–21

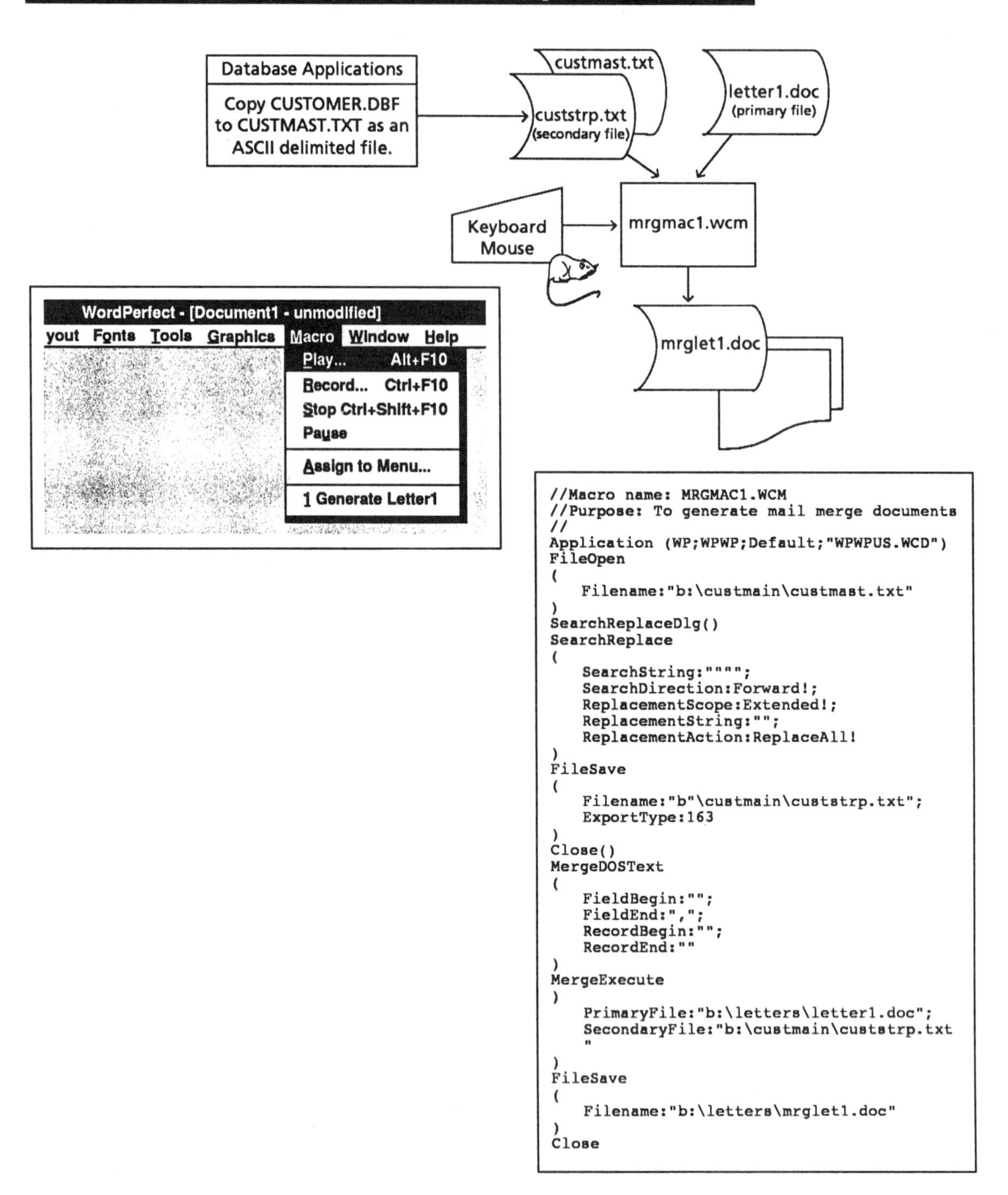

FIGURE 13–22 **Cornucopia Correspondence: Primary and Secondary Files**

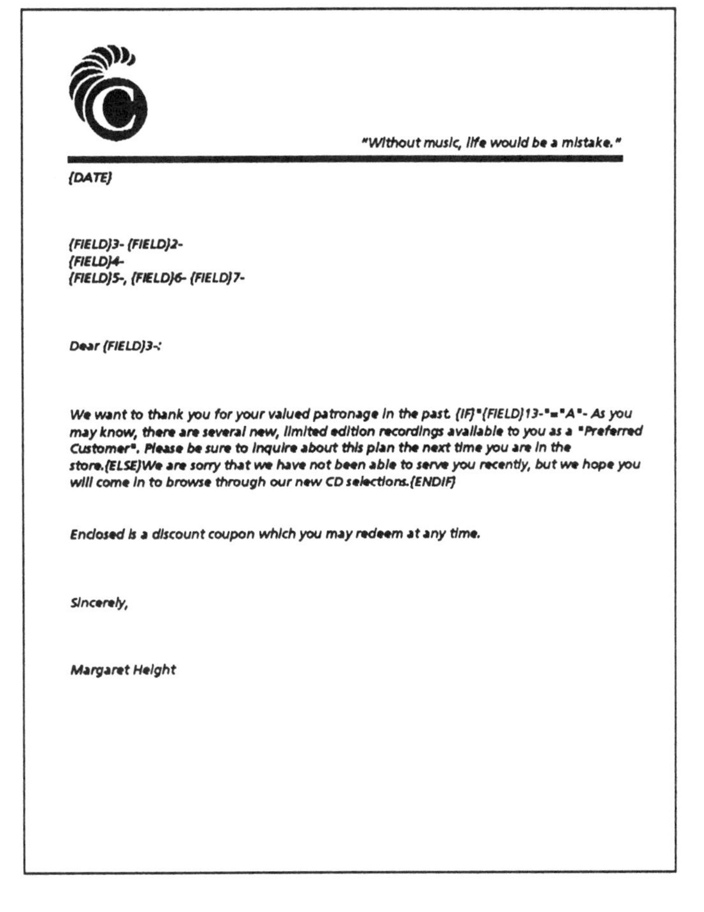

"Without music, life would be a mistake."

{DATE}

{FIELD}3- {FIELD}2-
{FIELD}4-
{FIELD}5-, {FIELD}6- {FIELD}7-

Dear {FIELD}3-:

We want to thank you for your valued patronage in the past. {IF}"{FIELD}13-"="A"- As you may know, there are several new, limited edition recordings available to you as a "Preferred Customer". Please be sure to inquire about this plan the next time you are in the store.{ELSE}We are sorry that we have not been able to serve you recently, but we hope you will come in to browse through our new CD selections.{ENDIF}

Enclosed is a discount coupon which you may redeem at any time.

Sincerely,

Margaret Height

Custmast.txt (ASCII delimited)

```
1111,"JOHNSON","PHIL","4567 HOWARD ST.","EUREKA","CA","95501","000","707","443","1234","19920801","D"
1112,"ALTON","JUNE","234 FRANCIS CT.","ARCATA","CA","95512","000","707","883","4567","19920803","D"
1113,"KIRTLAND","MASON","P.O. BOX 654","COLETA","CA","95506","0654","707","768","5678","19920715","A"
1114,"MELIZE","GERGIO","56 OAK ST.","FORTUNA","CA","95540","000","707","725","1234","19920824","A"
1115,"SQUIRT","JULIE","4321 GOLDEN HEIGHTS","FORTUNA","CA","95540","000","707","725","8989","1992082","A"
1116,"WALKER","HARRY","1212 C. ST.","EUREKA","CA","95502","000","707","443","6767","19920912","A"
1117,"TEMPLETON","DAVID","4567 BEACON WAY","FIELDS LANDING","CA","95507","000","707","667","1212","19920915","A"
1118,"SMITH","GEORGE","6789 G. ST.","EUREKA","CA","95501","000","707","443","6789","19920828","D"
1119,"LOWE","FRANCIS","4567 WALTER BLVD.","ARCATA","CA","95508","000","707","883","1234","19920901","D"
1120,"BORDEN","RICHARD","123 ORCHARD","FIELDBROOK","CA","95523","000","707","123","1235","19930328","A"
1121,"HARRIS","DAVE","123 COLLEGE","EUREKA","CA","95501","707","445","6762","19930330","A"
```

Custstrp.txt (secondary file, with quotes removed)

```
1111,JOHNSON,PHIL,4567 HOWARD ST.,EUREKA,CA,95501,000,707,443,1234,19920801,D
1112,ALTON,JUNE,234 FRANCIS CT.,ARCATA,CA,95512,000,707,883,4567,19920803,D
1113,KIRTLAND,MASON,P.O. BOX 654,COLETA,CA,95506,0654,707,768,5678,19920715,A
1114,MELIZE,GERGIO,56 OAK ST.,FORTUNA,CA,95540,000,707,725,1234,19920824,A
1115,SQUIRT,JULIE,4321 GOLDEN HEIGHTS,FORTUNA,CA,95540,000,707,725,8989,1992082,A
1116,WALKER,HARRY,1212 C. ST.,EUREKA,CA,95502,000,707,443,6767,19920912,A
1117,TEMPLETON,DAVID,4567 BEACON WAY,FIELDS LANDING,CA,95507,000,707,667,1212,19920915,A
1118,SMITH,GEORGE,6789 G. ST.,EUREKA,CA,95501,000,707,443,6789,19920828,D
1119,LOWE,FRANCIS,4567 WALTER BLVD.,ARCATA,CA,95508,000,707,883,1234,19920901,D
1120,BORDEN,RICHARD,123 ORCHARD,FIELDBROOK,CA,95523,000,707,123,1235,19930328,A
1121,HARRIS,DAVE,123 COLLEGE,EUREKA,CA,95501,707,445,6762,19930330,A
```

FIGURE 13–23

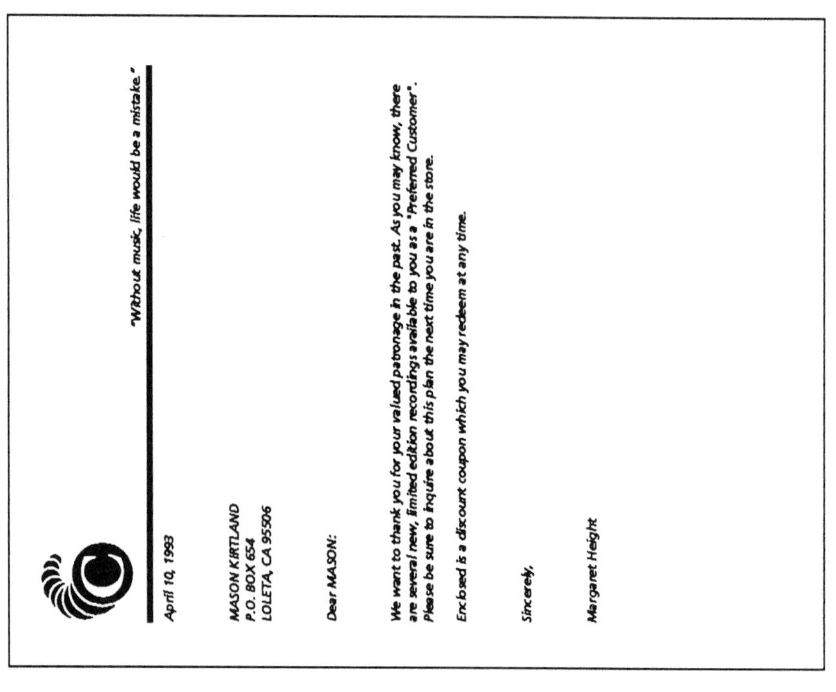

"Without music, life would be a mistake."

April 10, 1993

MASON KIRTLAND
P.O. BOX 654
LOLETA, CA 95506

Dear MASON:

We want to thank you for your valued patronage in the past. As you may know, there are several new, limited edition recordings available to you as a "Preferred Customer". Please be sure to inquire about this plan the next time you are in the store.

Enclosed is a discount coupon which you may redeem at any time.

Sincerely,

Margaret Height

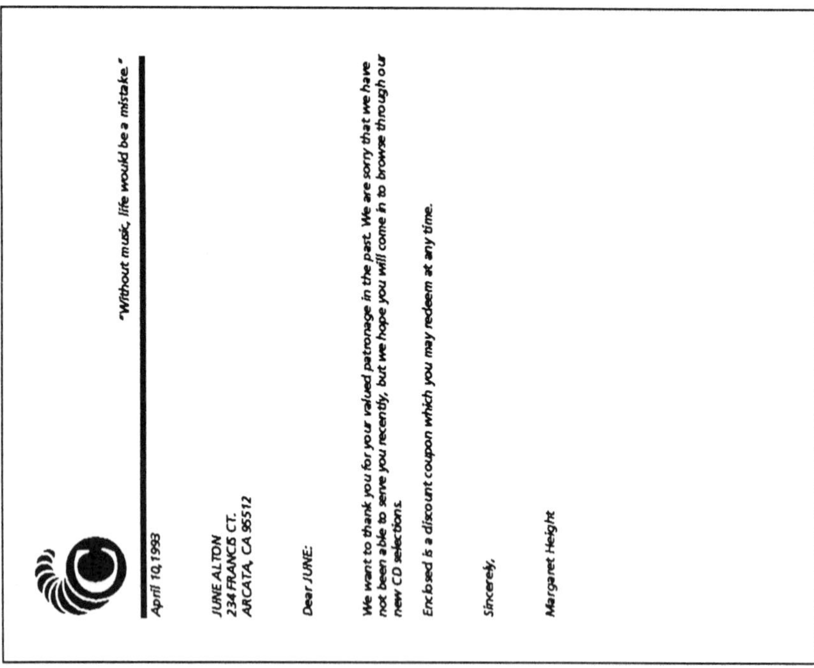

"Without music, life would be a mistake."

April 10, 1993

JUNE ALTON
234 FRANCIS CT.
ARCATA, CA 95512

Dear JUNE:

We want to thank you for your valued patronage in the past. We are sorry that we have not been able to serve you recently, but we hope you will come in to browse through our new CD selections.

Enclosed is a discount coupon which you may redeem at any time.

Sincerely,

Margaret Height

Cornucopia Sales/Profit Chart: System Flowchart **FIGURE 13–24**

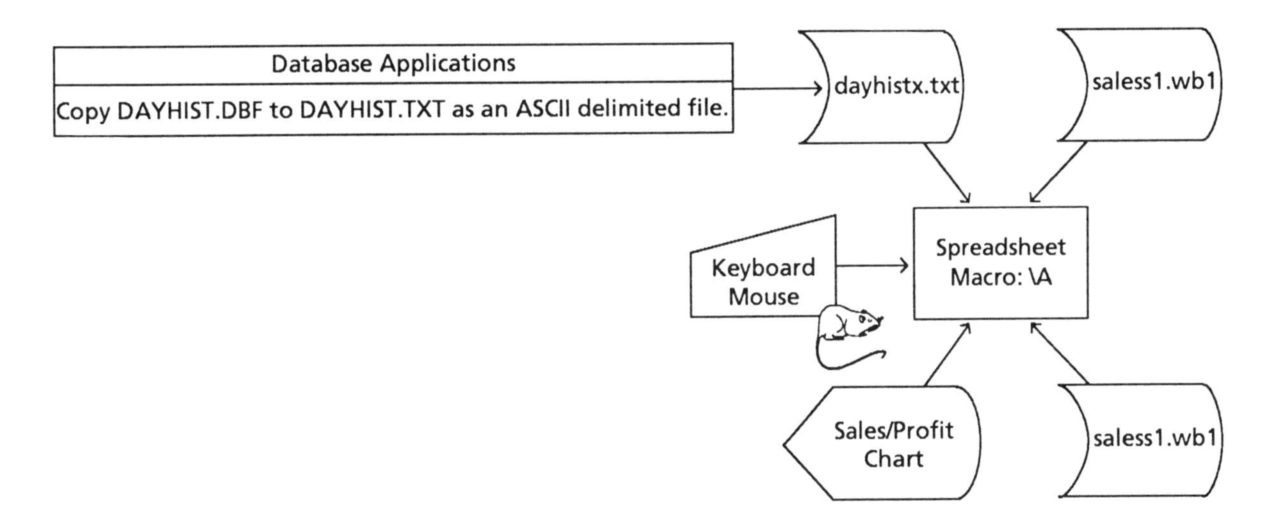

Spreadsheet Layout
SaleSS1.WB1

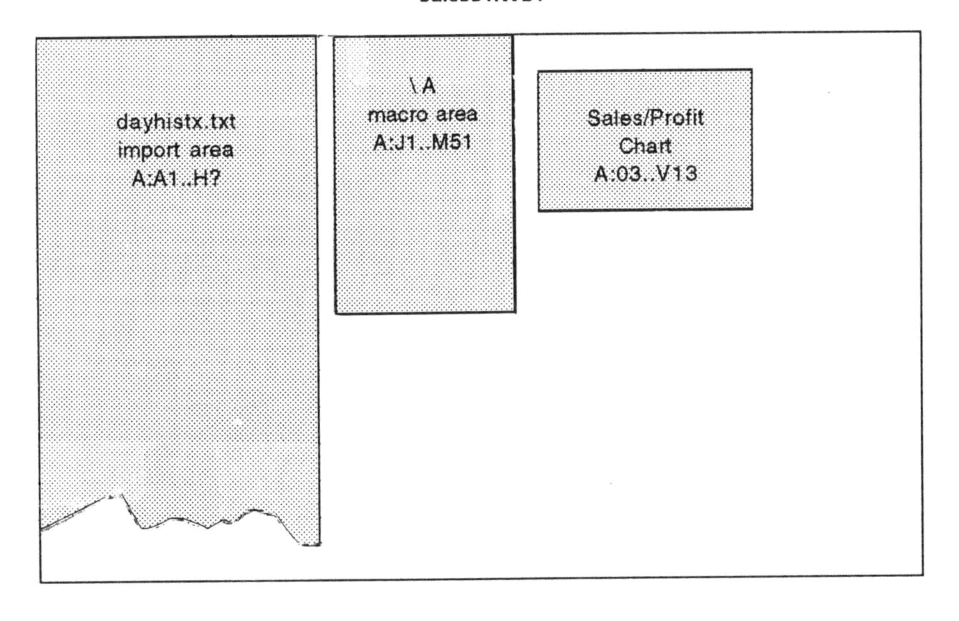

FIGURE 13–25 Cornucopia Sales/Profit Chart: Spreadsheet Macro

```
{;This is the SALESS1.WB1macro}

{SelectBlock A:A1..A1}
{FileImport"B:\SALETRAN\dayhistx.txt","Comma and ""Delimited File"}
{SelectBlock A:A1..A1}
{End}{Down}
{GetNumber"Enter the number of days to analyze?",number}
{For counter,1,number-1,1,_moveup}
{Up}
{For counter,1,6,1,_clearcell}
{Left 6}
{For counter,1,number+1,1,_movedown}
{For counter,1,6,1,_clearcell}
{Left 6}
{For counter,1,6,1,_sumlist}
{Left 6}
{Right}
{EditCopy}
{SelectBlock A:P10..P10}
{PasteSpecial Properties, Values,"",""}
{_gotosums}
{Right 2}
{EditCopy}
{SelectBlock A:P11..P11}
{PasteSpecial Properties, Values,"",""}
{_gotosums}
{Right 3}
{EditCopy}
{SelectBlock A:P12..P12}
{PasteSpecial Properties, Values,"",""}
{_gotosums}
{Right 4}
{EditCopy}
{SelectBlock A:Q10..Q10}
{PasteSpecial Properties, Values,"",""}
{_gotosums}
{Right 5}
{EditCopy}
{SelectBlock A:Q11..Q11}
{PasteSpecial Properties, Values,"",""}
{_gotosums}
{Right 6}
{EditCopy}
{SelectBlock A:Q12..Q12}
{PasteSpecial Properties, Values,"",""}
{SelectBlock AO1..O1}
{Message quit_msg,62,24,0}
{FileSave}
{SelectBlock A:J51..J51}

{;Close this file and EXIT Quattro Pro to return to the Application Window.}
```

```
counter      7
number       10

_moveup      {Up}

_movedown    {Down}

_sumlist     {Right}
             {Speedsum}

_clearcell   {Right}
             {EditClear}

_gotosums    {SelectBlock A:A1..A1}
             {End}{Down 2}

quit_msg     Press any key to QUIT.
```

```
C:\WINDOWS>type b:\saletran\dayhistx.txt
19930503,12,15,8,120,225,200,Y
19930504,10,34,12,100,545,295,Y
19930505,38,28,15,305,465,395,Y
19930506,31,27,18,300,425,465,Y
19930507,45,37,25,450,555,625,Y
19930508,52,41,22,480,620,560,Y
19930510,15,17,10,150,250,230,Y
19930511,12,15,12,65,235,290,Y
19930512,22,28,18,200,400,440,Y
19930513,37,32,23,340,495,580,Y
19930514,35,33,27,350,520,675,Y
19930515,40,35,30,400,525,750,Y
```

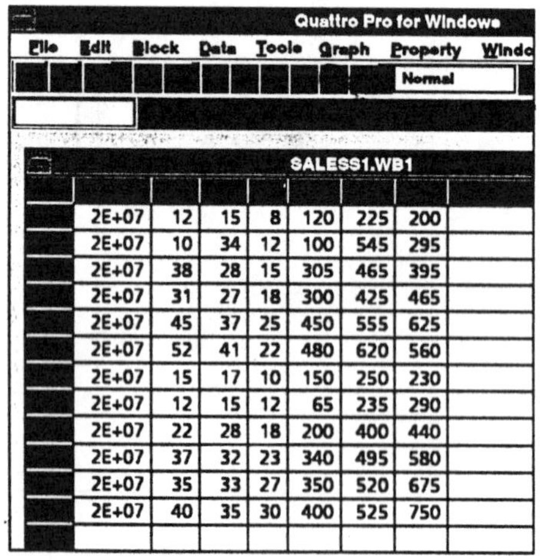

Quattro Pro for Windows

File Edit Block Data Tools Graph Property Windo

Normal

SALESS1.WB1

2E+07	12	15	8	120	225	200	
2E+07	10	34	12	100	545	295	
2E+07	38	28	15	305	465	395	
2E+07	31	27	18	300	425	465	
2E+07	45	37	25	450	555	625	
2E+07	52	41	22	480	620	560	
2E+07	15	17	10	150	250	230	
2E+07	12	15	12	65	235	290	
2E+07	22	28	18	200	400	440	
2E+07	37	32	23	340	495	580	
2E+07	35	33	27	350	520	675	
2E+07	40	35	30	400	525	750	

Cornucopia Sales/Profit Chart: Screen Report

FIGURE 13–26

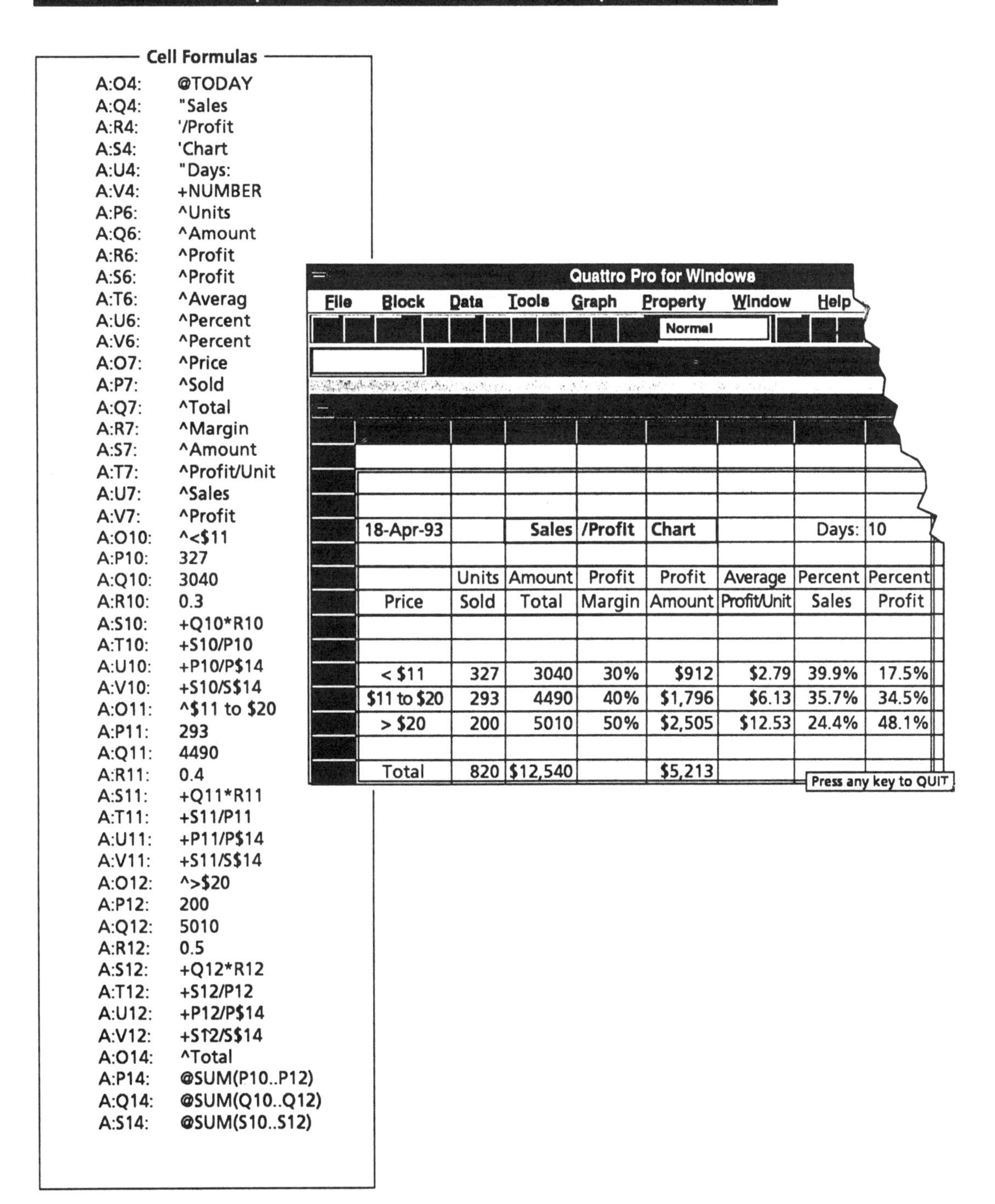

Cell Formulas

```
A:O4:   @TODAY
A:Q4:   "Sales
A:R4:   '/Profit
A:S4:   'Chart
A:U4:   "Days:
A:V4:   +NUMBER
A:P6:   ^Units
A:Q6:   ^Amount
A:R6:   ^Profit
A:S6:   ^Profit
A:T6:   ^Averag
A:U6:   ^Percent
A:V6:   ^Percent
A:O7:   ^Price
A:P7:   ^Sold
A:Q7:   ^Total
A:R7:   ^Margin
A:S7:   ^Amount
A:T7:   ^Profit/Unit
A:U7:   ^Sales
A:V7:   ^Profit
A:O10:  ^<$11
A:P10:  327
A:Q10:  3040
A:R10:  0.3
A:S10:  +Q10*R10
A:T10:  +S10/P10
A:U10:  +P10/P$14
A:V10:  +S10/S$14
A:O11:  ^$11 to $20
A:P11:  293
A:Q11:  4490
A:R11:  0.4
A:S11:  +Q11*R11
A:T11:  +S11/P11
A:U11:  +P11/P$14
A:V11:  +S11/S$14
A:O12:  ^>$20
A:P12:  200
A:Q12:  5010
A:R12:  0.5
A:S12:  +Q12*R12
A:T12:  +S12/P12
A:U12:  +P12/P$14
A:V12:  +S12/S$14
A:O14:  ^Total
A:P14:  @SUM(P10..P12)
A:Q14:  @SUM(Q10..Q12)
A:S14:  @SUM(S10..S12)
```


18-Apr-93		Sales	/Profit	Chart		Days:	10
	Units	Amount	Profit	Profit	Average	Percent	Percent
Price	Sold	Total	Margin	Amount	Profit/Unit	Sales	Profit
< $11	327	3040	30%	$912	$2.79	39.9%	17.5%
$11 to $20	293	4490	40%	$1,796	$6.13	35.7%	34.5%
> $20	200	5010	50%	$2,505	$12.53	24.4%	48.1%
Total	820	$12,540		$5,213			

Object-Oriented Output

Figure 13-27 illustrates a Paradox for Windows screen form designed for the Cornucopia project. It is composed of 45 objects, each with a set of properties and associated methods. Figure 13-28 identifies all 45 objects in an object tree, as well as the underlying data model that shows the one-to-many file relationship.

FIGURE 13–27 Cornucopia Object-Oriented Sales History Report

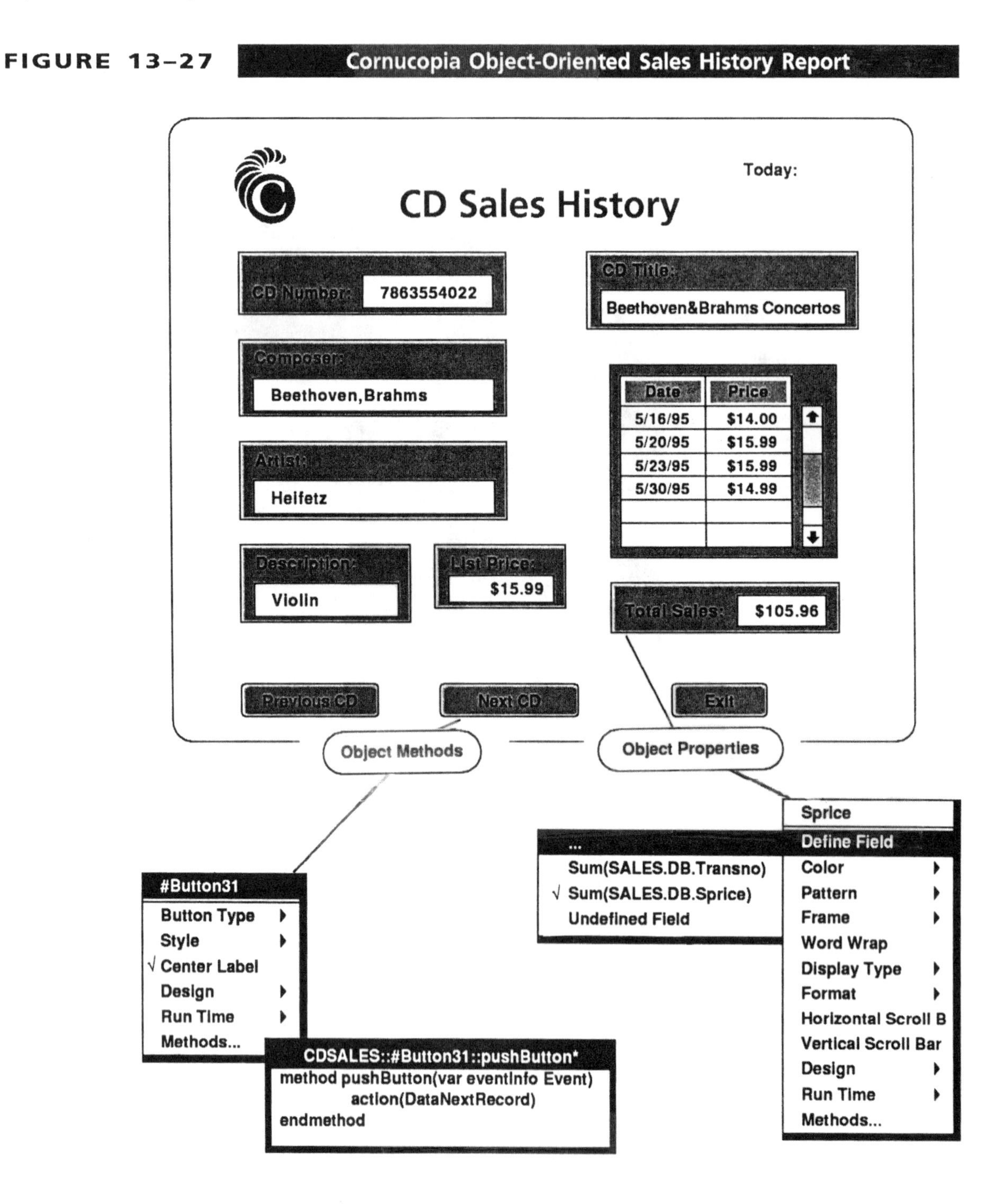

A careful examination of the object tree reveals the rich variety of object types available to the analyst-programmer. Can you identify the bitmap, line, and text objects that exist independently of other objects? Which object provides a base for the sales table objects? What object is a calculated field? What would you expect to happen when the user clicks the button objects? If the scroll bar associated with the sales table object is *not* an object, what is it? Finally, what does the structure of the object tree imply about object inheritance and modularity?

Cornucopia Sales History Object Tree FIGURE 13–28

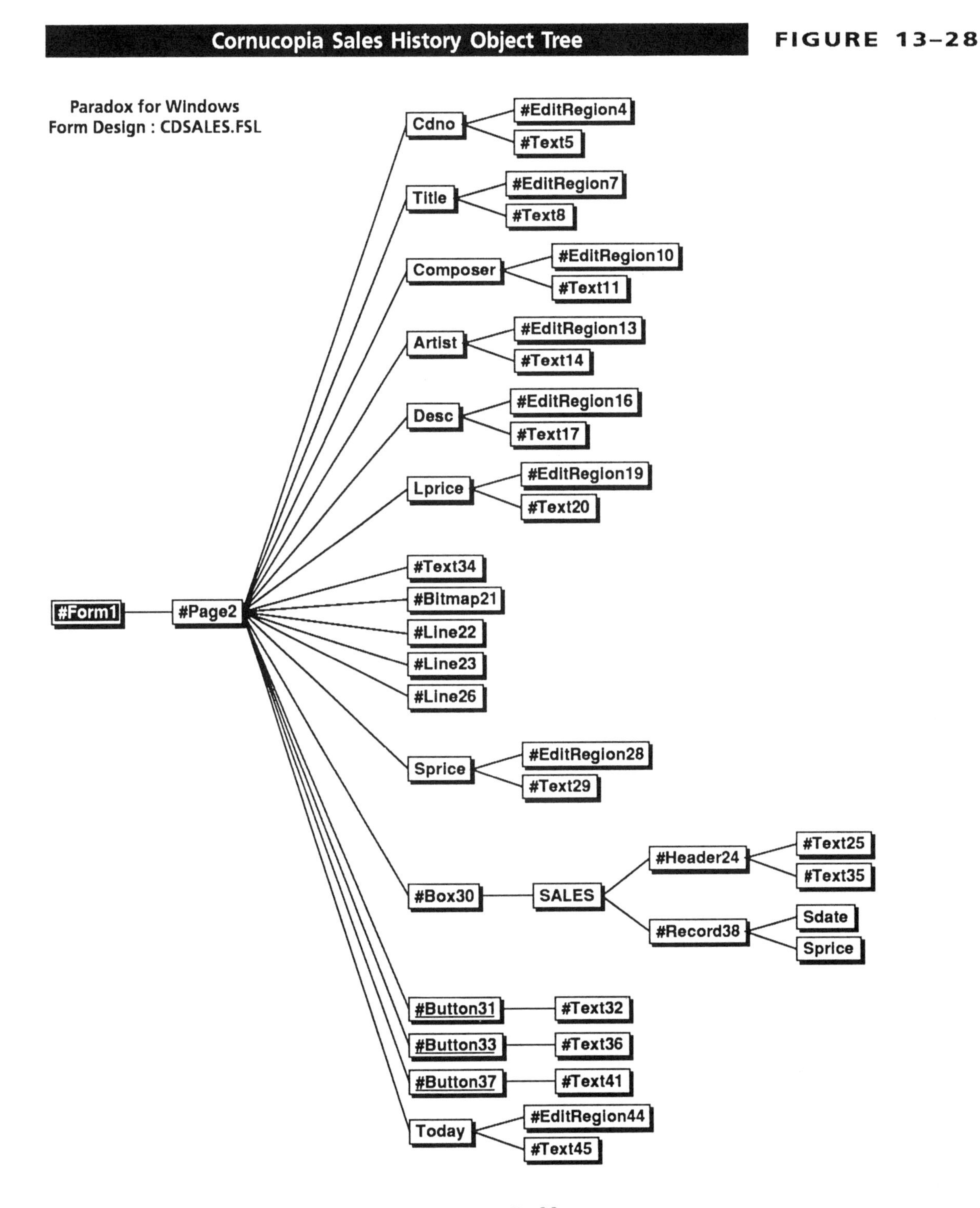

Time and Money

The analyst charges 20 hours to the development phase to cover the work performed during the last week. This pushes actual labor charges over the estimate by a whopping 17 hours ($850)! Still, the hardware and software savings amount to more than $1,000, so in the aggregate, the budget is okay. But, there can be no more outrageous overages can be allowed. The good news: Development is estimated tobe 60 percent complete, which puts us back on schedule. The lesson: It is better to spend the time to catch up with the schedule now, rather than to keep slipping behind and endangering the on-time delivery of the finished product.

FIGURE 13–29 Cornucopia Project Status—Week 10

Date: Cornucopia Project Status A/O Week: 10

Activity	% Comp.	Status	1	2	3	4	5	6	7	8	9	0	1	2	3	4	5	6	Total
Analysis -Est	100%		4	5	5	4													18
Actuals	100%	ok	2	4	2	3													11
Design -Est	100%					4	4	5	5	4									22
Actuals	95%	ok			1	3	8	5	7	6									30
Develop -Est	60%									4	4	5	5	4					22
Actuals	60%	ok								1	8	20							29
Impl. -Est	0%													4	5	5	5	5	24
Actuals	0%	ok																	0
Total -Est			4	5	5	8	4	5	5	8	4	5	5	8	5	5	5	5	86
Actuals			2	4	3	6	8	5	7	7	8	20	0	0	0	0	0	0	70
Contract	100%	ok	C																
Proj. Prelim.	100%	ok			C														
Design Rev.	100%	ok								C									
Proto. Rev.	0%	ok											S						
Train & Del.	0%	ok															S		
Final Rpt.	0%	ok																S	
			1	2	3	4	5	6	7	8	9	0	1	2	3	4	5	6	

Chapter 14
System Environments

The system resource requirements specifications (from Figure 11-9) include a mainstream PC microcomputer system. Nothing in the SDLC process suggests that the system will require a network solution at this time, although the possibility of a future upgrade to use telecommunications to help maintain the CD database has been mentioned. Also, the owner wants the analysts to discuss how the system could eventually be modified to a point-of-sale system. Therefore, the analysts choose to install the information system as a new program group within the Windows environment.

The System Interface

The original system interface appears as a series of simple menus in Figure 8-15. However, the final design, as presented in Chapter 13, involves three major horizontal software applications: word processing, spreadsheet, and database management. The database portion consists of several subsystems, which do correspond more closely to the initial menu design. The word processing and spreadsheet portions, although driven by macros, do not offer a menu interface.

The entire system takes advantage of the Windows Program Manager interfacing mechanism. A new Cornucopia program group contains three program items, each of which activates a different horizontal application that is customized for this information system. The user can alternatively use the standard Program Manager interface to launch the horizontal products independent of the information system. Figure 14-15 shows the sequence followed to activate the CD Maintenance subsystem. Figure 14-16 shows the Program Item Properties for the items in the Cornucopia group.

Notice that in the database application a batch file is used to launch dBASE IV and the mainmenu.prg. This is provided as an illustration of how a batch file can be used rather than the command line parameter. As illustrated in Figure 13-21, the mail merge operation is executed from the macro menu inside WordPerfect. Finally, the spreadsheet application macro is specified directly in the command line parameter.

The Prototype Review Session

Given the close association between the user and the analyst during the project activities to date, the prototype review session may seem to be an unnecessary formality. This was not the case in the Cornucopia project. After a rather routine presentation of the prototyping methodology, the user interrupted the session with a question about the point-of-sale options to the system design. She was particularly interested in the possibility of mechanizing the inventory control procedures of the enterprise.

The analysts were not really surprised by this because they had noticed the owner's computer confidence level and information system product appetite growing with each joint application design session. To the analysts' credit, they did not immediately reject the suggestion. Rather, they commented that such a major change at this point in the project might endanger the delivery date and cause significant budget overages. Further, they agreed to pursue the matter and return with a careful analysis of this suggestion within a few days.

Sample Cornucopia Access Sequence

FIGURE 14–15

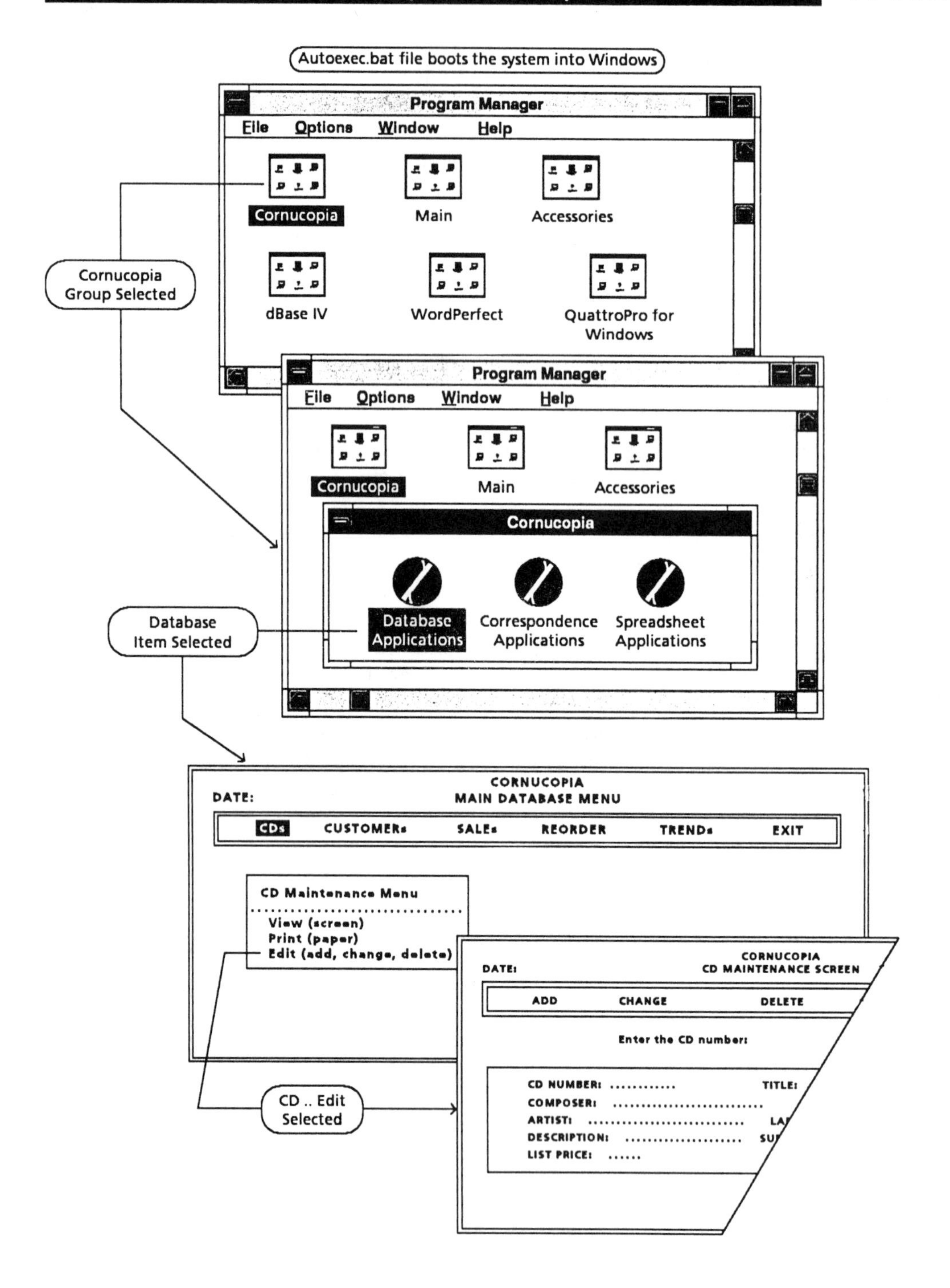

FIGURE 14–16 Cornucopia Program Item Properties

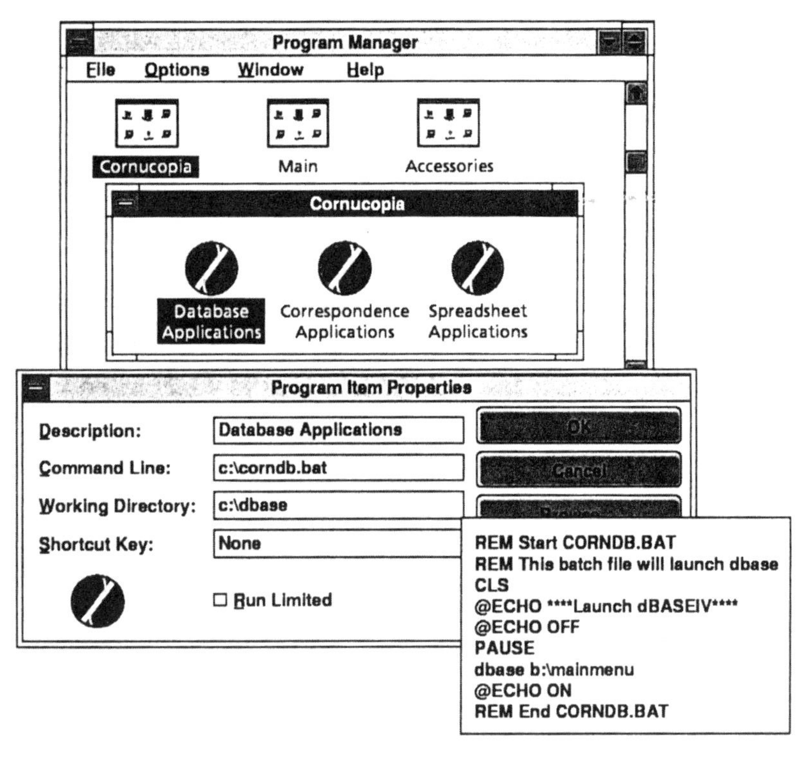

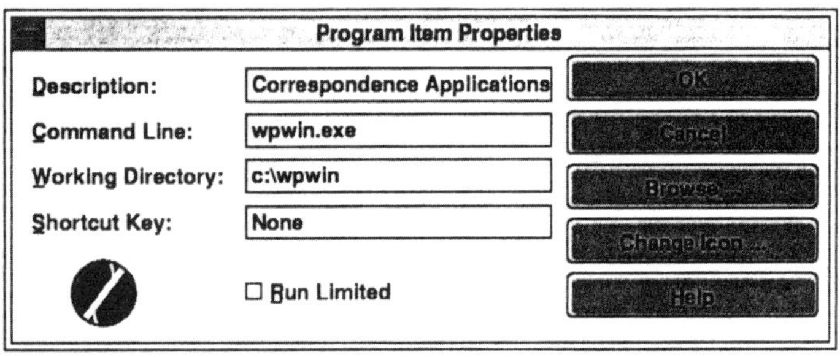

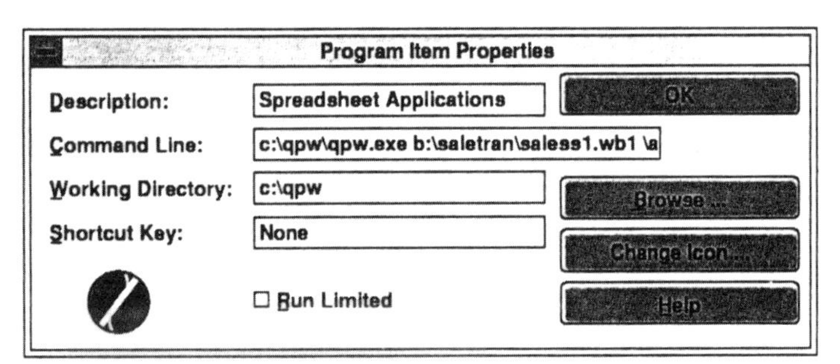

Time and Money

The analysts report 5 hours of work on the development phase of the project, bringing the estimated completion to 70 percent. As the Project Budget for Week 11 (Figure 14-17) indicates, the labor overages have been offset by savings in hardware and software expenses.

FIGURE 14–17 **Cornucopia Program Budget—Week 11**

Date: Cornucopia Budget A/O Wk: 11

	Week1	Week2	Week3	Week4	Week5	Week6	Week7	Week8	Week9	Week10	Week11	Week12	Week13	Week14	Week15	Week16	Total
Estimates																	
Hardware						2500	500	500	500								4000
Software						1000	250	250									1500
Labor	200	250	250	400	200	250	250	400	200	250	250	400	250	250	250	250	4300
Total	200	250	250	400	200	3750	1000	1150	700	250	250	400	250	250	250	250	9800
Actuals																	
Hardware						1995	595	495									3085
Software						495	149	289	235								1168
Labor	100	200	150	300	400	250	350	350	400	1000	250						3750
Total	100	200	150	300	400	2740	1094	1134	635	1000	250	0	0	0	0	0	8003
Weekly +/–																	
Hardware	0	0	0	0	0	505	-95	5	500	0	0	0	0	0	0	0	915
Software	0	0	0	0	0	505	101	-39	-235	0	0	0	0	0	0	0	332
Labor	100	50	100	100	-200	0	-100	50	-200	-750	0	0	0	0	0	0	-850
Total	100	50	100	100	-200	1010	-94	16	65	-750	0	0	0	0	0	0	397
Cumm. +/–																	
Hardware	0	0	0	0	0	505	410	415	915	915	915	0	0	0	0	0	
Software	0	0	0	0	0	505	606	567	332	332	332	0	0	0	0	0	
Labor	100	150	250	350	150	150	50	100	-100	-850	-850	0	0	0	0	0	
Total	100	150	250	350	150	1160	1066	1082	1147	397	397	0	0	0	0	0	

Chapter 15
System Testing

Cornucopia's design and development testing sequence is patterned after the plan illustrated in Figure 15-2. At this point in the sample project, testing has progressed through integrated module testing and the operating system interface. The next set of system tests should involve all five components of the system.

The Testing Sequence

Figure 15-9 cross references the exact sequence of tests with the appropriate illustrations in previous Cornucopia material. The system flowchart (from Figure 12-12) provides an excellent overview of all the system programs detailed in the figure.

The various subsystems were successfully developed and tested in the modular fashion described earlier. However, when the mainmenu program was tested, system failures occurred as each of the subsystems was called. Figure 15-10 documents one such test. After mainmenu calls custmain, custmain attempts to "clear" the screen and then "clear all," which is the offending code (line 5).

At first, the analyst-programmer was troubled by this error because these modules had executed successfully during the program module testing activities. A careful reading of the dBASE IV reference manual reveals that this command attempts to close all open menus, which is inconsistent with line 65 of mainmenu where we "activate menu mmain." Therefore, this error occurs only during integrated module testing. The solution is to remove the "clear all" command from custmain. This action does not affect the system. In fact, you may have noticed that several code segments are identical in these two programs. These segments were added to solve screen management problems that occurred during module testing, which explains how the "clear all" code was added in the first place.

Time and Money

The analysts report 6 hours to development activities, bringing the completion up to 90 percent. In addition, they report 4 hours to the implementation activities, with a 5 percent completion. Figure 15-11 presents the project status as of Week 12.

The recent user request to investigate the possibility of changing the system design to incorporate point-of-sale hardware and software and an inventory control subsystem is answered, in part, by the budget report, which shows only a modest budget surplus ($400) to date. The status report reminds the user that only four weeks remain before the system's scheduled completion date. Together with the analyst's quick estimate that the changes will require about $1,000 in hardware and software and at least 15 labor hours ($750), the user agrees to wait for the final report, which promises to include a more complete analysis of the proposed upgrade.

FIGURE 15–9 Cornucopia Testing Sequence

1. **Syntax Testing:**
   Continues throughout all testing activities.

2. **User Interface Testing:**
   Prototype Menus: (Figure 8–15) These menus are only crude shells; therefore, they were never tested *per se.*
   Prototype Master File Maintenance Screen Forms: (Figures 8–16 and 8–17) Sample master files were created before the lower portion of these forms could be built. They were tested to satisfy the visual requirements of the user.
   Prototype Master File Screen Form: (Figure 8–18) The multiple record display on the lower portion of the form required a special modification of the original screen-painted form. This program (Figure 13–19) was initially tested with random data. Later, when a program reference to the existing CD master file was added (i.e., referential integrity), testing was more disciplined.

3. **Program Module Testing:**
   Customer Maintenance Subsystem Modules: (Figure 13–13) The "Blankcus," "Custform," and "Browcust" modules were developed and tested in conjunction with the prototype development. The "Custadd," "Custchg," and "Custdel" modules required custom programming, as illustrated in Figures 13–14 through 13–16. These modules were tested with the same sample master file data used for the prototype. Once the Customer master file modules were debugged, they were copied and slightly modified to yield the CD master file maintenance modules.
   Sales Transaction Subsystem Modules: (Figure 13–17) The "Blanksal" module was developed and tested for the prototype. The "Sales" module (Figure 13–19) required much more custom programming than did "Salesum" or "Upsales," but a sample sales summary file ("Dayhist") was created to test these two modules.
   Correspondence Mail Merge Macro: (Figures 13–21 through 13–23) The primary file was created and the secondary file was converted from the sample dBase Customer master file. The macro testing was a chore because WordPerfect does not easily accommodate dBase files.
   Trending Spreadsheet Macro: (Figures 13–24 and 13–26) The dBase Sales History file was converted so that the macro could be tested.

4. **Integrated Module Testing:**
   Customer Maintenance Subsystem Menu: (Figure 13–14) The "Custmain" module ties the subsystem together. Although most of this module was built and tested at the same time as the other subsystem modules, it could not be completed until all the other modules were debugged. The CD Maintenance subsystem menu is a virtual clone of this module.
   Database Applications Main Menu: (Figure 13–12) The "Mainmenu" module ties all the database applications together under one menu. Two of the menu options (Reorder and Trends) were left as stubs so that the integrated testing could proceed. Note the following discussion concerning one testing failure.

5. **System Testing:**
   Cornucopia Program Group: (Figure 14–15) The three major application areas are represented in this group. Each application is called a "program item" in the Windows environment. Testing at the system level was made by the existing Windows facilities for creating new program groups and items.

Cornucopia Integrated Module Testing Failure

FIGURE 15–10

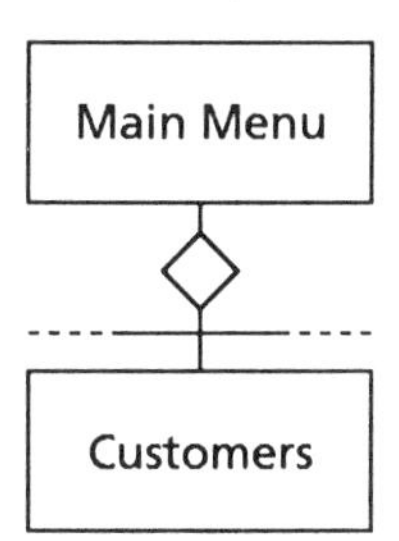

(1) Main Menu calls the Customers module.

```
*This is the main menu program.
*MAINMENU.PRG

clear
clear all
set talk off
set status off
set scoreboard off
@ 1,1 to 21,77 double
@ 2,35 say "CORNUCOPIA"
@ 3,4 say "DATE:"
@ 3,9 say date [ ]
@ 3,31 say "MAIN DATABASE MENU"

define menu main
   define pad mp1 of mmain prompt "CDs" at 5,9
   define pad mp2 of mmain prompt "CUSTOMERs" at 5,16
   :
activate menu mmain
   :
```

```
*This is the customer maintenance menu program.
*CUSTMAIN.PRG

clear
clear all
@ 1,1 to 21,77 double
@ 2,35 say "CORNUCOPIA"
@ 3,4 say "DATE:"
@ 3,9 say date [ ]
@ 3,27 say "CUSTOMER MAINTENANCE SCREEN"
:
@ 3,27 clear to 3,67
@ 3,31 say "MAIN DATABASE MENU"
@ 5,8 clear to 5,67
@ 5,9 say "CDs"
@ 5,16 say "CUSTOMERs"
:
```

(2) When Main Menu calls Customers, the system fails with the following message and screen display.

```
Cannot clear menu in use
clear all
** At line 5 in file custmain.prg, procedure CUSTMAIN
   from line 65 in file mainmenu.prg, procedure MAINMENU
```

FIGURE 15–11 Cornucopia Project Status—Week 12

Date: **Cornucopia Project Status** **A/O Week: 12**

Activity	% Comp.	Status	1	2	3	4	5	6	7	8	9	0	1	2	3	4	5	6	Total
Analysis -Est	100%		4	5	5	4													18
Actuals	100%	ok	2	4	2	3													11
Design -Est	100%					4	4	5	5	4									22
Actuals	95%	ok			1	3	8	5	7	6									30
Develop -Est	100%									4	4	5	5	4					22
Actuals	90%	ok								1	8	20	5	6					40
Impl. -Est	15%													4	5	5	5	5	24
Actuals	5%	ok												2					2
Total -Est			4	5	5	8	4	5	5	8	4	5	5	8	5	5	5	5	86
Actuals			2	4	3	6	8	5	7	7	8	20	5	8	0	0	0	0	83
Contract	100%	ok	C																
Proj. Prelim.	100%	ok			C														
Design Rev.	100%	ok								C									
Proto. Rev.	100%	ok											C						
Train & Del.	0%	ok															S		
Final Rpt.	0%	ok																S	
			1	2	3	4	5	6	7	8	9	0	1	2	3	4	5	6	

Chapter 16
System Documentation and Training

Cornucopia is a good example of a small enterprise that will have only two or three people accessing the computer information system, one of whom is the owner. At this time, the owner chooses to be the person responsible for the overall well-being of the computer and the information system. The sales clerks will need to know how to work with the master file maintenance and the transaction processing subsystems.

All these people participated in at least a portion of the prototyping and testing activities, but the owner has the most experience with the system so far. Fortunately, one of the sales clerks recently completed a computer literacy class at the local community college.

Training Activities

The training schedule appears in Figure 16-11. The sessions occur during a two-week period, with four of the contracted twelve hours (see the Project Contact in Chapter 3) reserved for two follow-up visits after the system is fully implemented. All the initial training takes place after closing. The follow-up sessions are scheduled for normal business hours that are historically not too busy. All sessions are conducted on-site, which means that the computer will be moved back and forth from the development location to the business premises until the system is fully implemented.

Product Documentation

Several important documents are prepared for the implementation activities. The training manual and procedures manual are used during the training sessions. The reference manual is primarily used when the system breaks down or for future analysts involved in a system upgrade. The user also receives the hardware and horizontal software manuals that come with those products.

Figure 16-12 shows the procedures manual entry describing the Reorder Report menu option. The analysts developed a standard format that includes the issue or revision date and the procedures manual section-reference, all highlighted by easily recognized content headings. One obvious drawback to printed manuals is that the amount of human intervention required to print and distribute documentation materials. Assembling product documentation into three-ring binders, with stylized section lettering and dividers, makes it easier to incorporate individual page versions to accommodate information system changes. However, any paper-based documentation system is subject to some haphazard page handling. Manual updates may become an increasingly expensive burden. An alternative to printed material is computer-based documentation.

Cornucopia Training Schedule

FIGURE 16–11

Session 1: Tuesday, 5:30–7:30 p.m., for all employees

1. Information System Overview
2. Demonstration of the Windows Interface
3. Demonstration of the Cornucopia Interface
4. Break
5. Hands-On Exercise on the Interfaces
6. Description of the Procedures Manual

Session 2: Wednesday, 5:30–7:30 p.m., for all employees

1. Hands-On Review Exercise
2. Demonstration of the MF Maintenance Subsystem
3. Hands-On Exercise on the Maintenance Subsystem
4. Break
5. Demonstration of the Sales Transaction Subsystem
6. Hands-On Exercise on the Sales Transaction Subsystem

Session 3: Tuesday, 5:30–7:30 p.m., for all employees

1. Hands-On Review Exercise
2. Demonstration of the 4GL Software Tutorials
3. Hands-On Problem-Solving Exercise
4. Break
5. Description of Emergency and Security Procedures
6. Review of Procedures Manual

Session 4: Wednesday, 5:30–7:30 p.m., for the owner

1. Demonstration of the Correspondence Subsystem
2. Demonstration of the Sales Trending Subsystem
3. Demonstration of the Reordering Subsystem
4. Hands-On Exercise on the Above Subsystems
5. Break
6. Overview of the System Hardware

FIGURE 16–12 Cornucopia Sample Procedures Manual Entry

: Reorder Report

Procedures Manual

Description: The reorder report lists the compact disks sold since the previous reorder report was generated.

Use: The reorder report is used to keep track of the disks that are sold and then reordered. Not all sold items will be replaced, which accounts for some blank "Ordered . . . Received" entries. You may use the comments field to note your intention to order the product at a later time or to discontinue the item, etc.

For those items that you do order, the report provides space for you to record the date you placed the order and the date the order is received.

Access: To generate a reorder report you must follow these instructions.

1. Activate the Database Applications option from the Cornucopia program group.

2. Choose the Reorder option from the Main Database Menu.

Operating Instructions : Menu Descriptions

M&M Section 2.2.5 Issued/Revised: 05/19/95

Time and Money

It is not always easy to differentiate the activities performed in one category from another category. This is true for the Cornucopia project team. This period they report 6 hours towards the development activities, increasing the percentage complete to 95 percent, and 4 hours toward the implementation activities, moving this phase to 40 percent complete. Figure 16-13 presents the status report as of Week 13. Notice that the total actual labor hours for development is double the estimate. Clearly, this would jeopardize the overall budget estimate if it were not for the reduced hardware and software prices that provide the offset. This experience may explain why some analysts prefer to inflate the budget estimates in the beginning. However, this is not a recommended approach. A more productive practice is to evaluate the differences between estimates and actuals in terms of unexpected project complexity, resource ordering and delivery delays, personnel inexperience, and so on. In other words, the analyst's ability to accurately develop budget estimates will improve by carefully evaluating past experience and making corrective adjustments.

Cornucopia Project Status—Week 13 **FIGURE 16–13**

Date: **Cornucopia Project Status** **A/O Week: 13**

Activity	% Comp.	Status	1	2	3	4	5	6	7	8	9	0	1	2	3	4	5	6	Total
Analysis -Est	100%		4	5	5	4													18
Actuals	100%	ok	2	4	2	3													11
Design -Est	100%					4	4	5	5	4									22
Actuals	95%	ok			1	3	8	5	7	6									30
Develop -Est	100%									4	4	5	5	4					22
Actuals	95%	ok								1	8	20	5	6	4				44
Impl. -Est	40%													4	5	5	5	5	24
Actuals	40%	ok												2	4				6
Total -Est			4	5	5	8	4	5	5	8	4	5	5	8	5	5	5	5	86
Actuals			2	4	3	6	8	5	7	7	8	20	5	8	8	0	0	0	91
Contract	100%	ok	C																
Proj. Prelim.	100%	ok			C														
Design Rev.	100%	ok								C									
Proto. Rev.	100%	ok											C						
Train & Del.	0%	ok															S		
Final Rpt.	0%	ok																S	
			1	2	3	4	5	6	7	8	9	0	1	2	3	4	5	6	

Chapter 17
System Conversion

From the very beginning of this project, the analysts and the owner enjoyed a good working relationship. The analysts were sensitive to the owner's caution and reluctance to introduce too much change into her enterprise operations. The owner devoted the time required to answer questions, evaluate product designs, and participate in problem solving sessions. At the end of the project, these roles reverse to the extent that the user sees lots of possibilities for new system features, while the analysts argue that it might be better to let the new system operate as is for a while.

This type of analyst/user partnership is only one element of the *enhanced* SDLC. Horizontal software products now include features that make the technology itself more user friendly and approachable. This encourages the user to participate actively in the entire process. The new software also provides powerful design and development tools that allow the analyst to change information system products more easily than ever before. This encourages analysts to engage in collaborative efforts, which often produce lots of new ideas. Finally, the hardware can be scaled in size, cost, and complexity to fit the information needs of the enterprise, as well as the needs of the humans who must embrace this new work partner.

File Preparation

The Customer master file already exists as an ASCII file on the owner's old Apple II computer. This file is easily imported into Borland's dBASE IV 1.5 format. However, several new fields have been created (cno, edate, and status), and they need to be populated with data. The alphabetized file holds 74 records. The customer numbers (cno) are assigned from 1001 to 1074 consecutively. The effective date (edate) is uniformly set as the cutover date. The status (status) is initialized as active (A) for all 74 customers. The entire task is estimated to take not more than one hour of the analyst's time.

The CD master file does not exist. With more than 3,000 CDs in thecurrent inventory and thousands more available through catalog sales, this appears to be a considerable task -- especially given the potentially nonstandard values for several fields (e.g., there may be several artists and several pieces on a single disk). Further, almost all of the 12 hours of enterprise personnel time contracted for training and file creation is scheduled for training.

Through their vertical software research, the analysts discovered a company that provides current and past CD data on disk. The $500, one-time cost is pricey, but fortunately the project is currently about $200 under budget. The analysts decide to purchase the data disk and agree to pay the extra $300 from their fee. They estimate this will save about 35 hours of labor (3,000 disks at 30 to 45 seconds per disk).

Conversion Option

The owner agrees that a direct conversion is the best alternative for this project. It will certainly minimize the cost. The risk is low because most of the new system is entirely new, and, besides, the existing manual system can be easily resurrected. The owner shares the analysts' feeling that the system is not too complex and that the current personnel will welcome the computer into the workplace.

Project Review

The owner is very happy with the new system. However, she is eager to upgrade to a telecommunications system so that she can download her disk file updates. Also, the point-of-sale and inventory upgrade is still a priority. The analysts agree that these changes would greatly improve the information system, but they caution against making any decision without a careful study of the costs involved. They recommend that the owner document these requests on the new "Request for Services" form they have included in the maintenance section of the procedures manual.

Time and Money

With 5 hours reported to the implementation phase (60 percent complete) and 1 more hour reported to the development phase, the project is now 28 percent over its estimated labor charge (21 hours greater than the 76 hours estimated). Figure 17-8 presents the project status for Week 14.

The extraordinary labor charges are a concern. The fact that the overall budget is within the original estimate should not obscure the issue. The money saved on hardware and software purchases ($1247) could have been used to upgrade the hardware, purchase the point-of-sale system and subscribe to the on-line CD disk service. Alternatively, the savings could have simply reduced the total cost to the user. Further, if the actual labor hour data is a true indication of the effort required to do the work, then the analysts must consider this in all future estimates, thus making their work more expensive, and perhaps less competitive.

Cornucopia Project Status—Week 14

FIGURE 17–7

Date: Cornucopia Project Status A/O Week: 14

Activity		% Comp.	Status	1	2	3	4	5	6	7	8	9	0	1	2	3	4	5	6	Total
Analysis	-Est	100%		4	5	5	4													18
	Actuals	100%	ok	2	4	2	3													11
Design	-Est	100%					4	4	5	5	4									22
	Actuals	95%	ok			1	3	8	5	7	6									30
Develop	-Est	100%									4	4	5	5	4					22
	Actuals	95%	ok								1	8	20	5	6	4	1			45
Impl.	-Est	60%													4	5	5	5	5	24
	Actuals	60%	ok												2	4	5			11
Total	-Est			4	5	5	8	4	5	5	8	4	5	5	8	5	5	5	5	86
	Actuals			2	4	3	6	8	5	7	7	8	20	5	8	8	6	0	0	97
Contract		100%	ok	C																
Proj. Prelim.		100%	ok			C														
Design Rev.		100%	ok								C									
Proto. Rev.		100%	ok											C						
Train & Del.		0%	ok															S		
Final Rpt.		0%	ok																S	
				1	2	3	4	5	6	7	8	9	0	1	2	3	4	5	6	

Chapter 18
Information System Maintenance and Review

The project formally concludes soon after the system becomes operational and the initial flurry of cutover problems are resolved. However, this is not the end of the system's life cycle. Assuming that system maintenance and upgrade needs are attended to, the analysts expect this system to have a functional life of three to five years. The procedures manual specifies how the information system should be maintained to help guarantee this type of performance.

Programmed Information System Reviews

The idea of a periodic review of the information system seems too formal for the owner. While she appreciates the need to evaluate all of her enterprise operations, she feels that this can be accomplished by simply observing the day to day operations. Under no circumstances is she interested in forming a review team.

The analysts suggest that the project objectives itemized in the project contract might serve as a set of performance standards for a very informal review. They argue that it is important to establish some historical reference points for comparison during future reviews. Reluctantly, the owner agrees to schedule a review during the month following the first year anniversary of system implementation.

Information System Maintenance

The analysts offer the owner a $100-per-month maintenance contract that includes two call outs per month, plus one regularly scheduled service call per month. The owner rejects this offer on the basis of her recent experience with other small-enterprise computer users, who tell her that their computers never break down. Also, she feels that the thirty-day free call out service that goes with the normal product warranty is sufficient to get the bugs out of the system.

However, the owner's persistence about upgrading the system prompts the analysts to develop a plan (Figure 18-8) for introducing several changes into the system.

These cost estimates are based on the fact that the original system already includes a fax/modem, a bar code scanner, and the startup CD data disk purchased at the end of the implementation phase. Continued CD updates are available through a subscription service ($500 per year). Combined with some minor systems work ($300) that is required to interface the transmitted data to the existing database and update the system documentation, the total cost of this upgrade can be recouped within a few months by saving clerical labor hours. The point-of-sale upgrade requires the purchase of cash drawer hardware and software ($800), plus the analyst labor ($400) to incorporate the sales data into the existing sales subsystem. The inventory system is a natural extension of the point-of-sale upgrade, requiring analyst labor ($600) to add the user interfaces and complete the documentation revisions.

These changes are scheduled over a six month period of time for two reasons. First, this will minimize the disruptions to the enterprise. Second, the analysts are already committed to several other projects, leaving only pockets of time for new work. The owner acknowledges this reasoning and adds two more advantages. A well spaced series of changes allows her to spread out the upgrade expense and it guarantees a continued relationship with the analysts, at least for the next nine months.

Time and Money

The final project status report (Figure 18-9) and budget (Figure 18-10) reveal that the project was delivered on time and within the total cost estimate. As previously discussed, the labor cost category was grossly underestimated by almost 25 percent. This was offset by lower actual costs for hardware and software. In their attempt to explain this, the analysts argue that the ninth and tenth weeks of the project, during which they were involved with prototyping and programming activities, presented special problems. They agree to continue to search for new application development products that promise to reduce the programming burden.

FIGURE 18–9 Cornucopia Project Status—Week 16

Date: Cornucopia Project Status A/O Week: 16

Activity	% Comp.	Status	1	2	3	4	5	6	7	8	9	0	1	2	3	4	5	6	Total
Analysis -Est	100%		4	5	5	4													18
Actuals	100%	ok	2	4	2	3													11
Design -Est	100%					4	4	5	5	4									22
Actuals	100%	ok			1	3	8	5	7	6									30
Develop -Est	100%									4	4	5	5	4					22
Actuals	100%	ok								1	8	20	5	6	4	1			45
Impl. -Est	100%													4	5	5	5	5	24
Actuals	100%	ok												2	4	5	6	4	21
Total -Est			4	5	5	8	4	5	5	8	4	5	5	8	5	5	5	5	86
Actuals			2	4	3	6	8	5	7	7	8	20	5	8	8	6	6	4	107
Contract	100%	ok	C																
Proj. Prelim.	100%	ok			C														
Design Rev.	100%	ok								C									
Proto. Rev.	100%	ok											C						
Train & Del.	100%	ok															C		
Final Rpt.	100%	ok																C	
			1	2	3	4	5	6	7	8	9	0	1	2	3	4	5	6	

FIGURE 18–10 Cornucopia Budget—Week 16

Date: Cornucopia Budget A/O Wk: 16

	Week1	Week2	Week3	Week4	Week5	Week6	Week7	Week8
Estimates								
Hardware						2500	500	500
Software						1000	250	250
Labor	200	250	250	400	200	250	250	400
Total	200	250	250	400	200	3750	1000	1150
Actuals								
Hardware						1995	595	495
Software						495	149	289
Labor	100	200	150	300	400	250	350	350
Total	100	200	150	300	400	2740	1094	1134
Weekly +/–								
Hardware	0	0	0	0	0	505	-95	5
Software	0	0	0	0	0	505	101	-39
Labor	100	50	100	100	-200	0	-100	50
Total	100	50	100	100	-200	1010	-94	16
Cumm. +/–								
Hardware	0	0	0	0	0	505	410	415
Software	0	0	0	0	0	505	606	567
Labor	100	150	250	350	150	150	50	100
Total	100	150	250	350	150	1160	1066	1082

	Week9	Week10	Week11	Week12	Week13	Week14	Week15	Week16	Total
Estimates									
Hardware	500								4000
Software									1500
Labor	200	250	250	400	250	250	250	250	4300
Total	700	250	250	400	250	250	250	250	9800
Actuals									
Hardware									3085
Software	235								1168
Labor	400	1000	250	400	400	300	300	200	5350
Total	635	1000	250	400	400	300	300	200	9603
Weekly +/–									
Hardware	500	0	0	0	0	0	0	0	915
Software	-235	0	0	0	0	0	0	0	332
Labor	-200	-750	0	0	-150	-50	-50	50	-1050
Total	65	-750	0	0	-150	-50	-50	50	197
Cumm. +/–									
Hardware	915	915	915	915	915	915	915	915	
Software	332	332	332	332	332	332	332	332	
Labor	-100	-850	-850	-850	-1000	-1050	-1100	-1050	
Total	1147	397	397	397	247	197	147	197	